Little Masterpieces of Science

by Various

I0505801

PREFACE

To a good many of us the inventor is the true hero for he multiplies the working value of life. He performs an old task with new economy, as when he devises a mowing-machine to oust the scythe; or he creates a service wholly new, as when he bids a landscape depict itself on a photographic plate. He, and his twin brother, the discoverer, have eyes to read a lesson that Nature has held for ages under the undiscerning gaze of other men. Where an ordinary observer sees, or thinks he sees, diversity, a Franklin detects identity, as in the famous experiment here recounted which proves lightning to be one and the same with a charge of the Leyden jar. Of a later day than Franklin, advantaged therefor by new knowledge and better opportunities for experiment, stood Faraday, the founder of modern electric art. His work gave the world the dynamo and motor, the transmission of giant powers, almost without toll, for two hundred miles at a bound. It is, however, in the carriage of but trifling quantities of motion, just enough for signals, that electricity thus far has done its most telling work. Among the men who have created the electric telegraph Joseph Henry has a commanding place. A short account of what he did, told in his own words, is here presented. Then follows a narrative of the difficult task of laying the first Atlantic cables, a task long scouted as impossible: it is a story which proves how much science may be indebted to unfaltering courage, to faith in ultimate triumph.

To give speech the wings of electricity, to enable friends in Denver and New York to converse with one another, is a marvel which only familiarity places beyond the pale of miracle. Shortly after he perfected the telephone Professor Bell described the steps which led to its construction. That recital is here reprinted.

A recent wonder of electric art is its penetration by a photographic ray of substances until now called opaque. Professor Roentgen's account of how he wrought this feat forms one of the most stirring chapters in the history of science. Next follows an account of the telegraph as it dispenses with metallic conductors altogether, and trusts itself to that weightless ether which brings to the eye the luminous wave. To this succeeds a chapter which considers what electricity stands for as one of the supreme resources of human wit, a resource transcending even flame itself, bringing articulate speech and writing to new planes of facility and usefulness. It is shown that the rapidity with which during a single century electricity has been subdued for human service, illustrates that progress has leaps as well as deliberate steps, so that at last a gulf, all but infinite, divides man from his next of kin.

At this point we pause to recall our debt to the physical philosophy which underlies the calculations of the modern engineer. In such an experiment as that of Count Rumford we observe how the corner-stone was laid of the knowledge that heat is motion, and that motion under whatever guise, as light, electricity, or what not, is equally beyond creation or annihilation, however elusively it may glide from phase to phase and vanish from view. In the mastery of Flame for the superseding of muscle, of breeze and waterfall, the chief credit rests with James Watt, the inventor of the steam engine. Beside him stands George Stephenson, who devised the locomotive which by abridging space has lengthened life and added to its highest pleasures. Our volume closes by narrating the competition which decided that Stephenson's "Rocket" was much superior to its rivals, and thus opened a new chapter in the history of mankind.

GEORGE ILES.

CONTENTS

by an electric pulse. Induces currents in coils when the magnetism is varied in their iron or steel cores. Observes the lines of magnetic force as iron filings are magnetized. A magnetic bar moved in and out of a coil of wire excites electricity therein,--mechanical motion is converted into electricity. Generates a current by spinning a copper plate in a horizontal plane. 7

HENRY, JOSEPH

INVENTION OF THE ELECTRIC TELEGRAPH

Improves the electro-magnet of Sturgeon by insulating its wire with silk thread, and by disposing the wire in several coils instead of one. Experiments with a large electro-magnet excited by nine distinct coils. Uses a battery so powerful that electro-magnets are produced one hundred times more energetic than those of Sturgeon. Arranges a telegraphic circuit more than a mile long and at that distance sounds a bell by means of an electro-magnet. 23

ILES, GEORGE

THE FIRST ATLANTIC CABLES

Forerunners at New York and Dover. Gutta-percha the indispensable insulator. Wire is used to sheathe the cables. Cyrus W. Field's project for an Atlantic cable. The first cable fails. 1858 so does the second cable 1865. A triumph of courage, 1866. The highway smoothed for successors. Lessons of the cable. 37

BELL, ALEXANDER GRAHAM

THE INVENTION OF THE TELEPHONE

Indebted to his father's study of the vocal organs as they form sounds. Examines the Helmholtz method for the analysis and synthesis of vocal sounds. Suggests the electrical actuation of tuning-forks and the electrical transmission of their tones. Distinguishes intermittent, pulsatory and undulatory currents. Devises as his first articulating telephone a harp of steel rods thrown into vibration by electro-magnetism. Exhibits optically the vibrations of sound, using a preparation of a human ear: is struck by the efficiency of a slight aural membrane. Attaches a bit of clock spring to a piece of goldbeater's skin, speaks to it, an audible message is received at a distant and similar device. This contrivance improved is shown at the Centennial Exhibition, Philadelphia, 1876. At first the same kind of instrument transmitted and delivered, a message; soon two distinct instruments were invented for transmitting and for receiving. Extremely small magnets suffice. A single blade of grass forms a telephonic circuit. 57

DAM, H. J. W.

PHOTOGRAPHING THE UNSEEN

Roentgen indebted to the researches of Faraday, Clerk-Maxwell, Hertz, Lodge and Lenard. The human optic nerve is affected by a very small range in the waves that exist in the ether. Beyond the visible spectrum of common light are vibrations which have long been known as heat or as photographically active. Crookes in a vacuous bulb produced soft light from high tension electricity. Lenard found that rays from a Crookes' tube passed through substances opaque to common light. Roentgen extended these experiments and used the rays photographically, taking pictures of the bones of the hand through living flesh, and so on. 87

ILES, GEORGE

THE WIRELESS TELEGRAPH

What may follow upon electric induction. Telegraphy to a moving train. The Preece induction method; its limits. Marconi's system. His precursors, Hertz, Onesti, Branly and Lodge. The coherer and the vertical wire form the essence of the apparatus. Wireless telegraphy at sea. 109

ILES, GEORGE

ELECTRICITY, WHAT ITS MASTERY MEANS: WITH A REVIEW AND A PROSPECT

Electricity does all that fire ever did, does it better, and performs uncounted services impossible to flame. Its mastery means as great a forward stride as the subjugation of fire. A minor invention or discovery simply adds to human resources: a supreme conquest as of flame or electricity, is a multiplier and lifts art and science to a new plane. Growth is slow, flowering is rapid: progress at times is so quick of pace as virtually to become a leap. The mastery of electricity based on that of fire. Electricity vastly wider of range than heat: it is energy in its most available and desirable phase. The telegraph and the telephone contrasted with the signal fire. Electricity as the servant of mechanic and engineer. Household uses of the current. Electricity as an agent of research now examines Nature in fresh aspects. The investigator and the commercial exploiter render aid to one another. Social benefits of electricity, in telegraphy, in quick travel. The current should serve every city house. 125

RUMFORD, COUNT (BENJAMIN THOMPSON)

HEAT AND MOTION IDENTIFIED

Observes that in boring a cannon much heat is generated: the longer the boring lasts, the more heat is produced. He argues that since heat without limit may be thus produced by motion, heat must be motion. 155

STEPHENSON, GEORGE

THE "ROCKET" LOCOMOTIVE AND ITS VICTORY

Shall it be a system of stationary engines or locomotives? The two best practical engineers of the day are in favour of stationary engines. A test of locomotives is, however, proffered, and George Stephenson and his son, Robert, discuss how they may best build an engine to win the first prize. They adopt a steam blast to stimulate the draft of the furnace, and raise steam quickly in a boiler having twenty-five small fire-tubes of copper. The "Rocket" with a maximum speed of twenty-nine miles an hour distances its rivals. With its load of water its weight was but four and a quarter tons. 163

INVENTION AND DISCOVERY

FRANKLIN IDENTIFIES LIGHTNING WITH ELECTRICITY

[From Franklin's Works, edited in ten volumes by John Bigelow, Vol. I, pages

Dr. Stuber, the author of the first continuation of Franklin's life, gives this account of the electrical experiments of Franklin:--

"His observations he communicated, in a series of letters, to his friend Collinson, the first of which is dated March 28, 1747. In these he shows the power of points in drawing and throwing off the electrical matter, which had hitherto escaped the notice of electricians. He also made the grand discovery of a plus and minus, or of a positive and negative state of electricity. We give him the honour of this without hesitation; although the English have claimed it for their countryman, Dr. Watson. Watson's paper is dated January 21, 1748; Franklin's July 11, 1747, several months prior. Shortly after Franklin, from his principles of the plus and minus state, explained in a satisfactory manner the phenomena of the Leyden phial, first observed by Mr. Cuneus, or by Professor Muschenbroeck, of Leyden, which had much perplexed philosophers. He showed clearly that when charged the bottle contained no more electricity than before, but that as much was taken from one side as thrown on the other; and that to discharge it nothing was necessary but to produce a communication between the two sides by which the equilibrium might be restored, and that then no signs of electricity would remain. He afterwards demonstrated by experiments that the electricity did not reside in the coating as had been supposed, but in the pores of the glass itself. After the phial was charged he removed the coating, and found that upon applying a new coating the shock might still be received. In the year 1749, he first suggested his idea of explaining the phenomena of thunder gusts and of aurora borealis upon electric principles. He points out many particulars in which lightning and electricity agree; and he adduces many facts, and reasonings from facts, in support of his positions.

"In the same year he conceived the astonishingly bold and grand idea of ascertaining the truth of his doctrine by actually drawing down the lightning, by means of sharp pointed iron rods raised into the regions of the clouds. Even in this uncertain state his passion to be useful to mankind displayed itself in a powerful manner. Admitting the identity of electricity and lightning, and knowing the power of points in repelling bodies charged with electricity, and in conducting fires silently and imperceptibly, he suggested the idea of securing houses, ships and the like from being damaged by lightning, by erecting pointed rods that should rise some feet above the most elevated part, and descend some feet into the ground or water. The effect of these he concluded would be either to prevent a stroke by repelling the cloud beyond the striking distance or by drawing off the electrical fire which it contained; or, if they could not effect this they would at least conduct the electrical matter to the earth without any injury to the building.

"It was not until the summer of 1752 that he was enabled to complete his grand and unparalleled discovery by experiment. The plan which he had originally proposed was, to erect, on some high tower or elevated place, a sentry-box from which should rise a pointed iron rod, insulated by being fixed in a cake of resin. Electrified clouds passing over this would, he conceived, impart to it a portion of their electricity which would be rendered evident to the senses by sparks being emitted when a key, the knuckle, or other conductor, was presented to it. Philadelphia at this time afforded no opportunity of trying an experiment of this kind. While Franklin was waiting for the erection of a spire, it occurred to him that he might have more ready access to the region of clouds by means of a common kite. He prepared one by fastening two cross sticks to a silk handkerchief, which would not suffer so much from the rain as paper. To the upright stick was affixed an iron point. The string was, as usual, of hemp, except the lower end, which was silk. Where the hempen string terminated, a key was fastened. With this apparatus, on the appearance of a thundergust approaching, he went out into the commons, accompanied by his son, to whom alone he communicated his intentions, well knowing the ridicule which, too generally for the interest of science, awaits unsuccessful experiments in philosophy. He placed himself under a shed, to avoid the rain; his kite was raised, a thunder-cloud passed over it, no sign of electricity appeared. He almost despaired of success, when suddenly he observed the loose fibres of his string to move towards an erect position. He now presented his knuckle to the key and received a strong spark. How exquisite must his sensations have been at this moment! On his experiment depended the fate of his theory. If he succeeded, his name would rank high among those who had improved science; if he failed, he must inevitably be subjected to the derision of mankind, or, what is worse, their pity, as a well-meaning man, but a weak, silly projector. The anxiety with which he looked for the result of his experiment may easily be conceived. Doubts and despair had begun to prevail, when the fact was ascertained, in so clear a manner, that even the most incredulous could no longer withhold their assent. Repeated sparks were drawn from the key, a phial was charged, a shock given, and all the experiments made which are usually performed with electricity."

FARADAY'S DISCOVERIES LEADING UP TO THE ELECTRIC DYNAMO AND MOTOR

[Michael Faraday was for many years Professor of Natural Philosophy at the Royal Institution, London, where his researches did more to subdue electricity to the service of man than those of any other physicist who ever lived. "Faraday as a Discoverer," by Professor John Tyndall (his successor) depicts a mind of the rarest ability and a character of the utmost charm. This biography is published by D. Appleton & Co., New York: the extracts which follow are from the third chapter.]

In 1831 we have Faraday at the climax of his intellectual strength, forty years of age, stored with knowledge and full of original power. Through reading, lecturing, and experimenting, he had become thoroughly familiar with electrical science: he saw where light was needed and expansion possible. The phenomena of ordinary electric induction belonged, as it were, to the alphabet of his knowledge: he knew that under ordinary circumstances the presence of an electrified body was sufficient to excite, by induction, an unelectrified body. He knew that the wire which carried an electric current was an electrified body, and still that all attempts had failed to make it excite in other wires a state similar to its own.

What was the reason of this failure? Faraday never could work from the experiments of others, however clearly described. He knew well that from every experiment issues a kind of radiation, luminous, in different degrees to different minds, and he hardly trusted himself to reason upon an experiment that he had not seen. In the autumn of 1831 he began to repeat the experiments with electric currents, which, up to that time, had produced no positive result. And here, for the sake of younger inquirers, if not for the sake of us all, it is worth while to dwell for a moment on a power which Faraday possessed in an extraordinary degree. He united vast strength with perfect flexibility. His momentum was that of a river, which combines weight and directness with the ability to yield to the flexures of its bed. The intentness of his vision in any direction did not apparently diminish his power of perception in other directions; and when he attacked a subject, expecting results, he had the faculty of keeping his mind alert, so that results different from those which he expected should not escape him through pre-occupation.

He began his experiments "on the induction of electric currents" by composing a helix of two insulated wires, which were wound side by side round the same wooden cylinder. One of these wires he connected with a voltaic battery of ten cells, and the other with a sensitive galvanometer. When connection with the battery was made, and while the current flowed, no effect whatever was observed at the galvanometer. But he never accepted an experimental result, until he had applied to it the utmost power at his command. He raised his battery from ten cells to one hundred and twenty cells, but without avail. The current flowed calmly through the battery wire without producing, during its flow, any sensible result upon the galvanometer.

"During its flow," and this was the time when an effect was expected--but here Faraday's power of lateral vision, separating, as it were from the line of expectation, came into play--he noticed that a feeble movement of the needle always occurred at the moment when he made contact with the battery; that the needle would afterwards return to its former position and remain quietly there unaffected by the flowing current. At the moment, however, when the circuit was interrupted the needle again moved, and in a direction opposed to that observed on the completion of the circuit.

This result, and others of a similar kind, led him to the conclusion "that the battery current through the one wire did in reality induce a similar current through the other; but that it continued for an instant only, and partook more of the nature of the electric wave from a common Leyden jar than of the current from a voltaic battery." The momentary currents thus generated were called induced currents, while the current which generated them was called the inducing current. It was immediately proved that the current generated at making the circuit was always opposed in direction to its generator, while that developed on the rupture of the circuit coincided in direction with the inducing current. It appeared as if the current on its first rush through the primary wire sought a purchase in the secondary one, and, by a kind of kick, impelled backward through the latter an electric wave, which subsided as soon as the primary current was fully established.

Faraday, for a time, believed that the secondary wire, though quiescent when the primary current had been once established, was not in its natural condition, its return to that condition being declared by the current observed at breaking the circuit. He called this hypothetical state of the wire the electro-tonic state: he afterwards abandoned this hypothesis, but seemed to return to it in after life. The term electro-tonic is also preserved by Professor Du Bois Reymond to express a certain electric condition of the nerves, and Professor Clerk Maxwell has ably defined and illustrated the hypothesis in the Tenth Volume of the "Transactions of the Cambridge Philosophical Society."

The mere approach of a wire forming a closed curve to a second wire through which a voltaic current flowed was then shown by Faraday to be sufficient to arouse in the neutral wire an induced current, opposed in direction to the inducing current; the withdrawal of the wire also generated a current having the same direction as the inducing current; those currents existed only during the time of approach or withdrawal, and when neither the primary nor the secondary wire was in motion, no matter how close their proximity might be, no induced current was generated.

Faraday has been called a purely inductive philosopher. A great deal of nonsense is, I fear, uttered in this land of England about induction and deduction. Some profess to befriend the one, some the other, while the real vocation of an investigator, like Faraday, consists in the incessant marriage of both. He was at this time full of the theory of Amprere, and it cannot be doubted that numbers of his experiments were executed merely to test his deductions from that theory. Starting from the discovery of Oersted, the celebrated French philosopher had shown that all the phenomena of magnetism then known might be reduced to the mutual attractions and repulsions of electric currents. Magnetism had been produced from electricity, and Faraday, who all his life long entertained a strong belief in such reciprocal actions, now attempted to effect the evolution of electricity from magnetism. Round a welded iron ring he placed two distinct coils of covered wire, causing the coils to occupy opposite halves of the ring. Connecting the ends of one of the coils with a galvanometer, he found that the moment the ring was magnetized, by sending a current through the other coil, the galvanometer needle whirled round four or five times in succession. The action, as before, was that of a pulse, which vanished immediately. On interrupting the current, a whirl of the needle in the opposite direction occurred. It was only during the time of magnetization or demagnetization that these effects were produced. The induced currents declared a change of condition only, and they vanished the moment the act of magnetization or demagnetization was complete.

The effects obtained with the welded ring were also obtained with straight bars of iron. Whether the bars were magnetized by the electric current, or were excited by the contact of permanent steel magnets, induced currents were always generated during the rise, and during the subsidence of the magnetism. The use of iron was then abandoned, and the same effects were obtained by merely thrusting a permanent steel magnet into a coil of wire. A rush of electricity through the coil accompanied the insertion of the magnet; an equal rush in the opposite direction accompanied its withdrawal. The precision with which Faraday describes these results, and the completeness with which he defined the boundaries of his facts, are wonderful. The magnet, for example, must not be passed quite through the coil, but only half through, for if passed wholly through, the needle is stopped as by a blow, and then he shows how this blow results from a reversal of the electric wave in the helix. He next operated with the powerful permanent magnet of the Royal Society, and obtained with it, in an exalted degree, all the foregoing phenomena.

And now he turned the light of these discoveries upon the darkest physical phenomenon of that day. Arago had discovered in 1824, that a disk of non-magnetic metal had the power of bringing a vibrating magnetic needle suspended over it rapidly to rest; and that on causing the disk to rotate the magnetic needle rotated along with it. When both were quiescent, there was not the slightest measurable attraction or repulsion exerted between the needle and the disk; still when in motion the disk was competent to drag after it, not only a light needle, but a heavy magnet. The question had been probed and investigated with admirable skill by both Arago and Ampere, and Poisson had published a theoretic memoir on the subject; but no cause could be assigned for so extraordinary an action. It had also been examined in this country by two celebrated men, Mr. Babbage and Sir John Herschel; but it still remained a mystery. Faraday always recommended the suspension of judgment in cases of doubt. "I have always admired," he says, "the prudence and philosophical reserve shown by M. Arago in resisting the temptations to give a theory of the effect he had discovered, so long as he could not devise one which was perfect in its application, and in refusing to assent to the imperfect theories of others." Now, however, the time for theory had come. Faraday saw mentally the rotating disk, under the operation of the magnet, flooded with his induced currents, and from the known laws of interaction between currents and magnets he hoped to deduce the motion observed by Arago. That hope he realized, showing by actual experiment that when his disk rotated currents passed through it, their position and direction being such as must, in accordance with the established laws of electro-magnetic action, produce the observed rotation.

Introducing the edge of his disk between the poles of the large horseshoe magnet of the Royal Society, and connecting the axis and the edge of the disk, each by a wire with a galvanometer, he obtained, when the disk was turned round, a constant flow of electricity. The direction of the current was determined by the direction of the motion, the current being reversed when the rotation was reversed. He now states the law which rules the production of currents in both disks and wires, and in so doing uses, for the first time, a phrase which has since become famous. When iron filings are scattered over a magnet, the particles of iron arrange themselves in certain determined lines called magnetic curves. In 1831, Faraday for the first time called these curves "lines of magnetic force;" and he showed that to produce induced currents neither approach to nor withdrawal from a magnetic source, or centre, or pole, was essential, but that it was only necessary to cut appropriately the lines of magnetic force. Faraday's first paper on Magneto-electric Induction, which I have here endeavoured to condense, was read before the Royal Society on the 24th of November, 1831.

On January 12, 1832, he communicated to the Royal Society a second paper on "Terrestrial Magneto-electric Induction," which was chosen as the Bakerian Lecture for the year. He placed a bar of iron in a coil of wire, and lifting the bar into the direction of the dipping needle, he excited by this action a current in the coil. On reversing the bar, a current in the opposite direction rushed through the wire. The same effect was produced, when, on holding the helix in the line of dip, a bar of iron was thrust into it. Here, however, the earth acted on the coil through the intermediation of the bar of iron. He abandoned the bar and simply set a copper-plate spinning in a horizontal plane; he knew that the earth's lines of magnetic force then crossed the plate at an angle of about 70? When the plate spun round, the lines of force were intersected and induced currents generated, which produced their proper effect when carried from the plate to the galvanometer. "When the plate was in the magnetic meridian, or in any other plane coinciding with the magnetic dip, then its rotation produced no effect upon the galvanometer."

At the suggestion of a mind fruitful in suggestions of a profound and philosophic character--I mean that of Sir John Herschel--Mr. Barlow, of Woolwich, had experimented with a rotating iron shell. Mr. Christie had also performed an elaborate series of experiments on a rotating iron disk. Both of them had found that when in rotation the body exercised a peculiar action upon the magnetic needle, deflecting it in a manner which was not observed during quiescence; but neither of them was aware at the time of the agent which produced this extraordinary deflection. They ascribed it to some change in the magnetism of the iron shell and disk.

But Faraday at once saw that his induced currents must come into play here, and he immediately obtained them from an iron disk. With a hollow brass ball, moreover, he produced the effects obtained by Mr. Barlow. Iron was in no way necessary: the only condition of success was that the rotating body should be of a character to admit of the formation of currents in its substance: it must, in other words, be a conductor of electricity. The higher the conducting power the more copious were the currents. He now passes from his little brass globe to the globe of the earth. He plays like a magician with the earth's magnetism. He sees the invisible lines along which its magnetic action is exerted and sweeping his wand across these lines evokes this new power. Placing a simple loop of wire round a magnetic needle he bends its upper portion to the west: the north pole of the needle immediately swerves to the east: he bends his loop to the east, and the north poles moves to the west. Suspending a common bar magnet in a vertical position, he causes it to spin round its own axis. Its pole being connected with one end of a galvanometer wire, and its equator with the other end, electricity rushes round the galvanometer from the rotating magnet. He remarks upon the "singular independence" of the magnetism and the body of the magnet which carries it. The steel behaves as if it were isolated from its own magnetism.

And then his thoughts suddenly widen, and he asks himself whether the rotating earth does not generate induced currents as it turns round its axis from west to east. In his experiment with the twirling magnet the galvanometer wire remained at rest; one portion of the circuit was in motion relatively to another portion. But in the case of the twirling planet the galvanometer wire would necessarily be carried along with the earth; there would be no relative motion. What must be the consequence? Take the case of a telegraph wire with its two terminal plates dipped into the earth, and suppose the wire to lie in the magnetic meridian. The ground underneath the wire is influenced like the wire itself by the earth's rotation; if a current from south to north be generated in the wire, a similar current from south to north would be generated in the earth under the wire; these currents would run against the same terminal plates, and thus neutralize each other.

This inference appears inevitable, but his profound vision perceived its possible invalidity. He saw that it was at least possible that the difference of conducting power between the earth and the wire might give one an advantage over the other, and that thus a residual or differential current might be obtained. He combined wires of different materials, and caused them to act in opposition to each other, but found the combination ineffectual. The more copious flow in the better conductor was exactly counterbalanced by the resistance of the worst. Still, though experiment was thus emphatic, he would clear his mind of all discomfort by operating on the earth itself. He went to the round lake near Kensington Palace, and stretched four hundred and eighty feet of copper wire, north and south, over the lake, causing plates soldered to the wire at its ends to dip into the water. The copper wire was severed at the middle, and the severed ends connected with a galvanometer. No effect whatever was observed. But though quiescent water gave no effect, moving water might. He therefore worked at London Bridge for three days during the ebb and flow of the tide, but without any satisfactory result. Still he urges, "Theoretically it seems a necessary consequence, that where water is flowing there electric currents should be formed. If a line be imagined passing from Dover to Calais through the sea, and returning through the land, beneath the water, to Dover, it traces out a circuit of conducting matter one part of which, when the water moves up or down the channel, is cutting the magnetic curves of the earth, while the other is relatively at rest.... There is every reason to believe that currents do run in the general direction of the circuit described, either one way or the other, according as the passage of the waters is up or down the channel." This was written before the submarine cable was thought of, and he once informed me that actual observation upon that cable had been found to be in accordance with his theoretic deduction.

Three years subsequent to the publication of these researches, that is to say on January 29, 1835, Faraday read before the Royal Society a paper "On the influence by induction of an electric current upon itself." A shock and spark of a peculiar character had been observed by a young man named William Jenkin, who must have been a youth of some scientific promise, but who, as Faraday once informed me, was dissuaded by his own father from having anything to do with science. The investigation of the fact noticed by Mr. Jenkin led Faraday to the discovery of the extra current, or the current induced in the primary wire itself at the moments of making and breaking contact, the phenomena of which he described and illustrated in the beautiful and exhaustive paper referred to.

Seven and thirty years have passed since the discovery of magneto-electricity; but, if we except the extra current, until quite recently nothing of moment was added to the subject. Faraday entertained the opinion that the discoverer of a great law or principle had a right to the "spoils"--this was his term--arising from its illustration; and guided by the principle he had discovered, his wonderful mind, aided by his wonderful ten fingers, overran in a single autumn this vast domain, and hardly left behind him the shred of a fact to be gathered by his successors.

And here the question may arise in some minds, What is the use of it all? The answer is, that if man's intellectual nature thirsts for knowledge then knowledge is useful because it satisfies this thirst. If you demand practical ends, you must, I think, expand your definition of the term practical, and make it include all that elevates and enlightens the intellect, as well as all that ministers to the bodily health and comfort of men. Still, if needed, an answer of another kind might be given to the question "what is its use?" As far as electricity has been applied for medical purposes, it has been almost exclusively Faraday's electricity. You have noticed those lines of wire which cross the streets of London. It is Faraday's currents that speed from place to place through these wires. Approaching the point of Dungeness, the mariner sees an unusually brilliant light, and from the noble lighthouse of La He the same light flashes across the sea. These are Faraday's sparks exalted by suitable machinery to sun-like splendour. At the present moment the Board of Trade and the Brethren of the Trinity House, as well as the Commissioners of Northern Lights, are contemplating the introduction of the Magneto-electric Light at numerous points upon our coasts; and future generations will be able to refer to those guiding stars in answer to the question, what has been the practical use of the labours of Faraday? But I would again emphatically say, that his work needs no justification, and that if he had allowed his vision to be disturbed by considerations regarding the practical use of his discoveries, those discoveries would never have been made by him. "I have rather," he writes in 1831, "been desirous of discovering new facts and new relations dependent on magneto-electric induction, than of exalting the force of those already obtained; being assured that the latter would find their full development hereafter."

In 1817, when lecturing before a private society in London on the element chlorine, Faraday thus expresses himself with reference to this question of utility. "Before leaving this subject, I will point out the history of this substance as an answer to those who are in the habit of saying to every new fact, 'What is its use?' Dr. Franklin says to such, 'What is the use of an infant?' The answer of the experimentalist is, 'Endeavour to make it useful.' When Scheele discovered this substance, it appeared to have no use; it was in its infancy and useless state, but having grown up to maturity, witness its powers, and see what endeavours to make it useful have done."

PROFESSOR JOSEPH HENRY'S INVENTION OF THE ELECTRIC TELEGRAPH

[In 1855 the Regents of the Smithsonian Institution, Washington, D. C., at the instance of their secretary, Professor Joseph Henry, took evidence with respect to his claims as inventor of the electric telegraph. The essential paragraphs of Professor Henry's statement are taken from the Proceedings of the Board of Regents of the Smithsonian Institution, Washington, 1857.]

There are several forms of the electric telegraph; first, that in which frictional electricity has been proposed to produce sparks and motion of pith balls at a distance.

Second, that in which galvanism has been employed to produce signals by means of bubbles of gas from the decomposition of water.

Third, that in which electro-magnetism is the motive power to produce motion at a distance; and again, of the latter there are two kinds of telegraphs, those in which the intelligence is indicated by the motion of a magnetic needle, and those in which sounds and permanent signs are made by the attraction of an electro-magnet. The latter is the class to which Mr. Morse's invention belongs. The following is a brief exposition of the several steps which led to this form of the telegraph.

The first essential fact which rendered the electro-magnetic telegraph possible was discovered by Oersted, in the winter of 1819-'20. It is illustrated by figure 1, in which the magnetic needle is deflected by the action of a current of galvanism transmitted through the wire A B.

The second fact of importance, discovered in 1820, by Arago and Davy, is illustrated in Fig. 2. It consists in this, that while a current of galvanism is passing through a copper wire A B, it is magnetic, it attracts iron filings and not those of copper or brass, and is capable of developing magnetism in soft iron.

The next important discovery, also made in 1820, by Ampere, was that two wires through which galvanic currents are passing in the same direction attract, and in the opposite direction, repel, each other. On this fact Ampere founded his celebrated theory, that magnetism consists merely in the attraction of electrical currents revolving at right angles to the line joining the two poles of the magnet. The magnetization of a bar of steel or iron, according to this theory consists in establishing within the metal by induction a series of electrical currents, all revolving in the same direction at right angles to the axis or length of the bar.

It was this theory which led Arago, as he states, to adopt the method of magnetizing sewing needles and pieces of steel wire, shown in Fig. 3. This method consists in transmitting a current of electricity through a helix surrounding the needle or wire to be magnetised. For the purpose of insulation the needle was enclosed in a glass tube, and the several turns of the helix were at a distance from each other to insure the passage of electricity through the whole length of the wire, or, in other words, to prevent it from seeking a shorter passage by cutting across from one spire to another. The helix employed by Arago obviously approximates the arrangement required by the theory of Ampere, in order to develop by induction the magnetism of the iron. By an attentive perusal of the original account of the experiments of Arago, it will be seen that, properly speaking, he made no electro-magnet, as has been asserted by Morse and others; his experiments were confined to the magnetism of iron filings, to sewing needles and pieces of steel wire of the diameter of a millimetre, or of about the thickness of a small knitting needle.

Mr. Sturgeon, in 1825, made an important step in advance of the experiments of Arago, and produced what is properly known as the electro-magnet. He bent a piece of iron wire into the form of a horseshoe, covered it with varnish to insulate it, and surrounded it with a helix, of which the spires were at a distance. When a current of galvanism was passed through the helix from a small battery of a single cup the iron wire became magnetic, and continued so during the passage of the current. When the current was interrupted the magnetism disappeared, and thus was produced the first temporary soft iron magnet.

The electro-magnet of Sturgeon is shown in Fig. 4. By comparing Figs. 3 and 4 it will be seen that the helix employed by Sturgeon was of the same kind as that used by Arago; instead however, of a straight steel wire inclosed in a tube of glass, the former employed a bent wire of soft iron. The difference in the arrangement at first sight might appear to be small, but the difference in the results produced was important, since the temporary magnetism developed in the arrangement of Sturgeon was sufficient to support a weight of several pounds, and an instrument was thus produced of value in future research.

The next improvement was made by myself. After reading an account of the galvanometer of Schweigger, the idea occurred to me that a much nearer approximation to the requirements of the theory of Ampere could be attained by insulating the conducting wire itself, instead of the rod to be magnetized, and by covering the whole surface of the iron with a series of coils in close contact. This was effected by insulating a long wire with silk thread, and winding this around the rod of iron in close coils from one end to the other. The same principle was extended by employing a still longer insulated wire, and winding several strata of this over the first, care being taken to insure the insulation between each stratum by a covering of silk ribbon. By this arrangement the rod was surrounded by a compound helix formed of a long wire of many coils, instead of a single helix of a few coils, (Fig. 5).

In the arrangement of Arago and Sturgeon the several turns of wire were not precisely at right angles to the axis of the rod, as they should be, to produce the effect required by the theory, but slightly oblique, and therefore each tended to develop a separate magnetism not coincident with the axis of the bar. But in winding the wire over itself, the obliquity of the several turns compensated each other, and the resultant action was at right angles to the bar. The arrangement then introduced by myself was superior to those of Arago and Sturgeon, first in the greater multiplicity of turns of wire, and second in the better application of these turns to the development of magnetism. The power of the instrument with the same amount of galvanic force, was by this arrangement several times increased.

The maximum effect, however, with this arrangement and a single battery was not yet obtained. After a certain length of wire had been coiled upon the iron, the power diminished with a further increase of the number of turns. This was due to the increased resistance which the longer wire offered to the conduction of electricity. Two methods of improvement therefore suggested themselves. The first consisted, not in increasing the length of the coil, but in using a number of separate coils on the same piece of iron. By this arrangement the resistance to the conduction of the electricity was diminished and a greater quantity made to circulate around the iron from the same battery. The second method of producing a similar result consisted in increasing the number of elements of the battery, or, in other words, the projectile force of the electricity, which enabled it to pass through an increased number of turns of wire, and thus, by increasing the length of the wire, to develop the maximum power of the iron.

To test these principles on a larger scale, the experimental magnet was constructed, which is shown in Fig. 6. In this a number of compound helices were placed on the same bar, their ends left projecting, and so numbered that they could be all united into one long helix, or variously combined in sets of lesser length.

From a series of experiments with this and other magnets it was proved that, in order to produce the greatest amount of magnetism from a battery of a single cup, a number of helices is required; but when a compound battery is used, then one long wire must be employed, making many turns around the iron, the length of wire and consequently the number of turns being commensurate with the projectile power of the battery.

In describing the results of my experiments, the terms intensity and quantity magnets were introduced to avoid circumlocution, and were intended to be used merely in a technical sense. By the intensity magnet I designated a piece of soft iron, so surrounded with wire that its magnetic power could be called into operation by an intensity battery, and by a quantity magnet, a piece of iron so surrounded by a number of separate coils, that its magnetism could be fully developed by a quantity battery.

I was the first to point out this connection of the two kinds of the battery with the two forms of the magnet, in my paper in Silliman's Journal, January, 1831, and clearly to state that when magnetism was to be developed by means of a compound battery, one long coil was to be employed, and when the maximum effect was to be produced by a single battery, a number of single strands were to be used.

These steps in the advance of electro-magnetism, though small, were such as to interest and astonish the scientific world. With the same battery used by Mr. Sturgeon, at least a hundred times more magnetism was produced than could have been obtained by his experiment. The developments were considered at the time of much importance in a scientific point of view, and they subsequently furnished the means by which magneto-electricity, the phenomena of dia-magnetism, and the magnetic effects on polarized light were discovered. They gave rise to the various forms of electro-magnetic machines which have since exercised the ingenuity of inventors in every part of the world, and were of immediate applicability in the introduction of the magnet to telegraphic purposes. Neither the electro-magnet of Sturgeon nor any electro-magnet ever made previous to my investigations was applicable to transmitting power to a distance.

The principles I have developed were properly appreciated by the scientific mind of Dr. Gale, and applied by him to operate Mr. Morse's machine at a distance.

Previous to my investigations the means of developing magnetism in soft iron were imperfectly understood. The electro-magnet made by Sturgeon, and copied by Dana, of New York, was an imperfect quantity magnet, the feeble power of which was developed by a single battery. It was entirely inapplicable to a long circuit with an intensity battery, and no person possessing the requisite scientific knowledge, would have attempted to use it in that connection after reading my paper.

In sending a message to a distance, two circuits are employed, the first a long circuit through which the electricity is sent to the distant station to bring into action the second, a short one, in which is the local battery and magnet for working the machine. In order to give projectile force sufficient to send the power to a distance, it is necessary to use an intensity battery in the long circuit, and in connection with this, at the distant station, a magnet surrounded with many turns of one long wire must be employed to receive and multiply the effect of the current enfeebled by its transmission through the long conductor. In the local or short circuit either an intensity or a quantity magnet may be employed. If the first be used, then with it a compound battery will be required; and, therefore on account of the increased resistance due to the greater quantity of acid, a less amount of work will be performed by a given amount of material; and, consequently, though this arrangement is practicable it is by no means economical. In my original paper I state that the advantages of a greater conducting power, from using several wires in the quantity magnet, may, in a less degree, be obtained by substituting for them one large wire; but in this case, on account of the greater obliquity of the spires and other causes, the magnetic effect would be less. In accordance with these principles, the receiving magnet, or that which is introduced into the long circuit, consists of a horseshoe magnet surrounded with many hundred turns of a single long wire, and is operated with a battery of from twelve to twenty-four elements or more, while in the local circuit it is customary to employ a battery of one or two elements with a much thicker wire and fewer turns.

It will, I think, be evident to the impartial reader that these were improvements in the electro-magnet, which first rendered it adequate to the transmission of mechanical power to a distance; and had I omitted all allusion to the telegraph in my paper, the conscientious historian of science would have awarded me some credit, however small might have been the advance which I made. Arago and Sturgeon, in the accounts of their experiments, make no mention of the telegraph, and yet their names always have been and will be associated with the invention. I briefly, however, called attention to the fact of the applicability of my experiments to the construction of the telegraph; but not being familiar with the history of the attempts made in regard to this invention, I called it "Barlow's project," while I ought to have stated that Mr. Barlow's investigation merely tended to disprove the possibility of a telegraph.

I did not refer exclusively to the needle telegraph when, in my paper, I stated that the magnetic action of a current from a trough is at least not sensibly diminished by passing through a long wire. This is evident from the fact that the immediate experiment from which this deduction was made was by means of an electro-magnet and not by means of a needle galvanometer.

At the conclusion of the series of experiments which I described in Silliman's Journal, there were two applications of the electro-magnet in my mind: one the production of a machine to be moved by electro-magnetism, and the other the transmission of or calling into action power at a distance. The first was carried into execution in the construction of the machine described in Silliman's Journal, vol. xx, 1831, and for the purpose of experimenting in regard to the second, I arranged around one of the upper rooms in the Albany Academy a wire of more than a mile in length, through which I was enabled to make signals by sounding a bell, (Fig. 7.) The mechanical arrangement for effecting this object was simply a steel bar, permanently magnetized, of about ten inches in length, supported on a pivot, and placed with its north end between the two arms of a horseshoe magnet. When the latter was excited by the current, the end of the bar thus placed was attracted by one arm of the horseshoe, and repelled by the other, and was thus caused to move in a horizontal plane and its further extremity to strike a bell suitably adjusted.

I also devised a method of breaking a circuit, and thereby causing a large weight to fall. It was intended to illustrate the practicability of calling into action a great power at a distance capable of producing mechanical effects; but as a description of this was not printed, I do not place it in the same category with the experiments of which I published an account, or the facts which could be immediately deduced from my papers in Silliman's Journal.

From a careful investigation of the history of electro-magnetism in its connection with the telegraph, the following facts may be established:

1. Previous to my investigations the means of developing magnetism in soft iron were imperfectly understood, and the electro-magnet which then existed was inapplicable to the transmission of power to a distance.

2. I was the first to prove by actual experiment that, in order to develop magnetic power at a distance, a galvanic battery of intensity must be employed to project the current through the long conductor, and that a magnet surrounded by many turns of one long wire must be used to receive this current.

3. I was the first actually to magnetize a piece of iron at a distance, and to call attention to the fact of the applicability of my experiments to the telegraph.

4. I was the first to actually sound a bell at a distance by means of the electro-magnet.

5. The principles I had developed were applied by Dr. Gale to render Morse's machine effective at a distance.

THE FIRST ATLANTIC CABLES

GEORGE ILES

[From "Flame, Electricity and the Camera," copyright Doubleday, Page & Co., New York.]

Electric telegraphy on land has put a vast distance between itself and the mechanical signalling of Chapp? just as the scope and availability of the French invention are in high contrast with the rude signal fires of the primitive savage. As the first land telegraphs joined village to village, and city to city, the crossing of water came in as a minor incident; the wires were readily committed to the bridges which spanned streams of moderate width. Where a river or inlet was unbridged, or a channel was too wide for the roadway of the engineer, the question arose, May we lay an electric wire under water? With an ordinary land line, air serves as so good a non-conductor and insulator that as a rule cheap iron may be employed for the wire instead of expensive copper. In the quest for non-conductors suitable for immersion in rivers, channels, and the sea, obstacles of a stubborn kind were confronted. To overcome them demanded new materials, more refined instruments, and a complete revision of electrical philosophy.

As far back as 1795, Francisco Salva had recommended to the Academy of Sciences, Barcelona, the covering of subaqueous wires by resin, which is both impenetrable by water and a non-conductor of electricity. Insulators, indeed, of one kind and another, were common enough, but each of them was defective in some quality indispensable for success. Neither glass nor porcelain is flexible, and therefore to lay a continuous line of one or the other was out of the question. Resin and pitch were even more faulty, because extremely brittle and friable. What of such fibres as hemp or silk, if saturated with tar or some other good non-conductor? For very short distances under still water they served fairly well, but any exposure to a rocky beach with its chafing action, any rub by a passing anchor, was fatal to them. What the copper wire needed was a covering impervious to water, unchangeable in composition by time, tough of texture, and non-conducting in the highest degree. Fortunately all these properties are united in gutta-percha: they exist in nothing else known to art. Gutta-percha is the hardened juice of a large tree (Isonandra gutta) common in the Malay Archipelago; it is tough and strong, easily moulded when moderately heated. In comparison with copper it is but one 60,000,000,000,000,000,000th as conductive. As without gutta-percha there could be no ocean telegraphy, it is worth while recalling how it came within the purview of the electrical engineer.

In 1843 Jos?d'Almeida, a Portuguese engineer, presented to the Royal Asiatic Society, London, the first specimens of gutta-percha brought to Europe. A few months later, Dr. W. Montgomerie, a surgeon, gave other specimens to the Society of Arts, of London, which exhibited them; but it was four years before the chief characteristic of the gum was recognized. In 1847 Mr. S. T. Armstrong of New York, during a visit to London, inspected a pound or two of gutta-percha, and found it to be twice as good a non-conductor as glass. The next year, through his instrumentality, a cable covered with this new insulator was laid between New York and Jersey City; its success prompted Mr Armstrong to suggest that a similarly protected cable be submerged between America and Europe. Eighteen years of untiring effort, impeded by the errors inevitable to the pioneer, stood between the proposal and its fulfilment. In 1848 the Messrs. Siemens laid under water in the port of Kiel a wire covered with seamless gutta-percha, such as, beginning with 1847, they had employed for subterranean conductors. This particular wire was not used for telegraphy, but formed part of a submarine-mine system. In 1849 Mr. C. V. Walker laid an experimental line in the English Channel; he proved the possibility of signalling for two miles through a wire covered with gutta-percha, and so prepared the way for a venture which joined the shores of France and England.

In 1850 a cable twenty-five miles in length was laid from Dover to Calais, only to prove worthless from faulty insulation and the lack of armour against dragging anchors and fretting rocks. In 1851 the experiment was repeated with success. The conductor now was not a single wire of copper, but four wires, wound spirally, so as to combine strength with flexibility; these were covered with gutta-percha and surrounded with tarred hemp. As a means of imparting additional strength, ten iron wires were wound round the hemp--a feature which has been copied in every subsequent cable (Fig. 58). The engineers were fast learning the rigorous conditions of submarine telegraphy; in its essentials the Dover-Calais line continues to be the type of deep-sea cables to-day. The success of the wire laid across the British Channel incited other ventures of the kind. Many of them, through careless construction or unskilful laying, were utter failures. At last, in 1855, a submarine line 171 miles in length gave excellent service, as it united Varna with Constantinople; this was the greatest length of satisfactory cable until the submergence of an Atlantic line.

In 1854 Cyrus W. Field of New York opened a new chapter in electrical enterprise as he resolved to lay a cable between Ireland and Newfoundland, along the shortest line that joins Europe to America. He chose Valentia and Heart's Content, a little more than 1,600 miles apart, as his termini, and at once began to enlist the co-operation of his friends. Although an unfaltering enthusiast when once his great idea had possession of him, Mr. Field was a man of strong common sense. From first to last he went upon well-ascertained facts; when he failed he did so simply because other facts, which he could not possibly know, had to be disclosed by costly experience. Messrs.

Whitehouse and Bright, electricians to his company, were instructed to begin a preliminary series of experiments. They united a continuous stretch of wires laid beneath land and water for a distance of 2,000 miles, and found that through this extraordinary circuit they could transmit as many as four signals per second. They inferred that an Atlantic cable would offer but little more resistance, and would therefore be electrically workable and commercially lucrative.

In 1857 a cable was forthwith manufactured, divided in halves, and stowed in the holds of the Niagara of the United States navy, and the Agamemnon of the British fleet. The Niagara sailed from Ireland; the sister ship proceeded to Newfoundland, and was to meet her in mid-ocean. When the Niagara had run out 335 miles of her cable it snapped under a sudden increase of strain at the paying-out machinery; all attempts at recovery were unavailing, and the work for that year was abandoned. The next year it was resumed, a liberal supply of new cable having been manufactured to replace the lost section, and to meet any fresh emergency that might arise. A new plan of voyages was adopted: the vessels now sailed together to mid-sea, uniting there both portions of the cable; then one ship steamed off to Ireland, the other to the Newfoundland coast. Both reached their destinations on the same day, August 5, 1858, and, feeble and irregular though it was, an electric pulse for the first time now bore a message from hemisphere to hemisphere. After 732 despatches had passed through the wire it became silent forever. In one of these despatches from London, the War Office countermanded the departure of two regiments about to leave Canada for England, which saved an outlay of about $250,000. This widely quoted fact demonstrated with telling effect the value of cable telegraphy.

Now followed years of struggle which would have dismayed any less resolute soul than Mr. Field. The Civil War had broken out, with its perils to the Union, its alarms and anxieties for every American heart. But while battleships and cruisers were patrolling the coast from Maine to Florida, and regiments were marching through Washington on their way to battle, there was no remission of effort on the part of the great projector.

Indeed, in the misunderstandings which grew out of the war, and that at one time threatened international conflict, he plainly saw how a cable would have been a peace-maker. A single word of explanation through its wire, and angry feelings on both sides of the ocean would have been allayed at the time of the Trent affair. In this conviction he was confirmed by the English press; the London Times said: "We nearly went to war with America because we had no telegraph across the Atlantic." In 1859 the British government had appointed a committee of eminent engineers to inquire into the feasibility of an Atlantic telegraph, with a view to ascertaining what was wanting for success, and with the intention of adding to its original aid in case the enterprise were revived. In July, 1863, this committee presented a report entirely favourable in its terms, affirming "that a well-insulated cable, properly protected, of suitable specific gravity, made with care, tested under water throughout its progress with the best-known apparatus, and paid into the ocean with the most improved machinery, possesses every prospect of not only being successfully laid in the first instance, but may reasonably be relied upon to continue for many years in an efficient state for the transmission of signals."

Taking his stand upon this endorsement, Mr. Field now addressed himself to the task of raising the large sum needed to make and lay a new cable which should be so much better than the old ones as to reward its owners with triumph. He found his English friends willing to venture the capital required, and without further delay the manufacture of a new cable was taken in hand. In every detail the recommendations of the Scientific Committee were carried out to the letter, so that the cable of 1865 was incomparably superior to that of 1858. First, the central copper wire, which was the nerve along which the lightning was to run, was nearly three times larger than before. The old conductor was a strand consisting of seven fine wires, six laid around one, and weighed but 107 pounds to the mile. The new was composed of the same number of wires, but weighed 300 pounds to the mile. It was made of the finest copper obtainable.

To secure insulation, this conductor was first embedded in Chatterton's compound, a preparation impervious to water, and then covered with four layers of gutta-percha, which were laid on alternately with four thin layers of Chatterton's compound. The old cable had but three coatings of gutta-percha, with nothing between. Its entire insulation weighed but 261 pounds to the mile, while that of the new weighed 400 pounds.[1] The exterior wires, ten in number, were of Bessemer steel, each separately wound in pitch-soaked hemp yarn, the shore ends specially protected by thirty-six wires girdling the whole. Here was a combination of the tenacity of steel with much of the flexibility of rope. The insulation of the copper was so excellent as to exceed by a hundredfold that of the core of 1858--which, faulty though it was, had, nevertheless, sufficed for signals. So much inconvenience and risk had been encountered in dividing the task of cable-laying between two ships that this time it was decided to charter a single vessel, the Great Eastern, which, fortunately, was large enough to accommodate the cable in an unbroken length. Foilhommerum Bay, about six miles from Valentia, was selected as the new Irish terminus by the company. Although the most anxious care was exercised in every detail, yet, when 1,186 miles had been laid, the cable parted in 11,000 feet of water, and although thrice it was grappled and brought toward the surface, thrice it slipped off the grappling hooks and escaped to the ocean floor. Mr. Field was obliged to return to England and face as best he might the men whose capital lay at the bottom of the sea-- perchance as worthless as so much Atlantic ooze. With heroic persistence he argued that all difficulties would yield to a renewed attack. There must be redoubled precautions and vigilance never for a moment relaxed. Everything that deep-sea telegraphy has since accomplished was at that moment daylight clear to his prophetic view. Never has there been a more signal example of the power of enthusiasm to stir cold-blooded men of business; never has there been a more striking illustration of how much science may depend for success upon the intelligence and the courage of capital. Electricians might have gone on perfecting exquisite apparatus for ocean telegraphy, or indicated the weak points in the comparatively rude machinery which made and laid the cable, yet their exertions would have been wasted if men of wealth had not responded to Mr. Field's renewed appeal for help. Thrice these men had invested largely, and thrice disaster had pursued their ventures; nevertheless they had faith surviving all misfortunes for a fourth attempt.

   In 1866 a new company was organized, for two objects: first, to recover the cable lost the previous year and complete it to the American shore; second, to lay another beside it in a parallel course. The Great Eastern was again put in commission, and remodelled in accordance with the experience of her preceding voyage. This time the exterior wires of the cable were of galvanized iron, the better to resist corrosion. The paying-out machinery was reconstructed and greatly improved. On July 13, 1866, the huge steamer began running out her cable twenty-five miles north of the line struck out during the expedition of 1865; she arrived without mishap in Newfoundland on July 27, and electrical communication was re-established between America and Europe. The steamer now returned to the spot where she had lost the cable a few months before; after eighteen days' search it was brought to the deck in good order. Union was effected with the cable stowed in the tanks below, and the prow of the vessel was once more turned to Newfoundland. On September 8th this second cable was safely landed at Trinity Bay. Misfortunes now were at an end; the courage of Mr. Field knew victory at last; the highest honors of two continents were showered upon him.

'Tis not the grapes of Canaan that repay, But the high faith that failed not by the way.

What at first was as much a daring adventure as a business enterprise has now taken its place as a task no more out of the common than building a steamship, or rearing a cantilever bridge. Given its price, which will include too moderate a profit to betray any expectation of failure, and a responsible firm will contract to lay a cable across the Pacific itself. In the Atlantic lines the uniformly low temperature of the ocean floor (about 4?C.), and the great pressure of the superincumbent sea, co-operate in effecting an enormous enhancement both in the insulation and in the carrying capacity of the wire. As an example of recent work in ocean telegraphy let us glance at the cable laid in 1894, by the Commercial Cable Company of New York. It unites Cape Canso, on the northeastern coast of Nova Scotia, to Waterville, on the southwestern coast of Ireland. The central portion of this cable much resembles that of its predecessor in 1866. Its exterior armour of steel wires is much more elaborate. The first part of Fig. 59 shows the details of manufacture: the central copper core is covered with gutta-percha, then with jute, upon which the steel wires are spirally wound, followed by a strong outer covering. For the greatest depths at sea, type A is employed for a total length of 1,420 miles; the diameter of this part of the cable is seven-eighths of an inch. As the water lessens in depth the sheathing increases in size until the diameter of the cable becomes one and one-sixteenth inches for 152 miles, as type B. The cable now undergoes a third enlargement, and then its fourth and last proportions are presented as it touches the shore, for a distance of one and three-quarter miles, where type C has a diameter of two and one-half inches. The weights of material used in this cable are: copper wire, 495 tons; gutta-percha, 315 tons; jute yarn, 575 tons; steel wire, 3,000 tons; compound and tar, 1,075 tons; total, 5,460 tons. The telegraph-ship Faraday, specially designed for cable-laying, accomplished the work without mishap.

Electrical science owes much to the Atlantic cables, in particular to the first of them. At the very beginning it banished the idea that electricity as it passes through metallic conductors has anything like its velocity through free space. It was soon found, as Professor Mendenhall says, "that it is no more correct to assign a definite velocity to electricity than to a river. As the rate of flow of a river is determined by the character of its bed, its gradient, and other circumstances, so the velocity of an electric current is found to depend on the conditions under which the flow takes place."[2] Mile for mile the original Atlantic cable had twenty times the retarding effect of a good aerial line; the best recent cables reduce this figure by nearly one-half.

In an extreme form, this slowing down reminds us of the obstruction of light as it enters the atmosphere of the earth, of the further impediment which the rays encounter if they pass from the air into the sea. In the main the causes which hinder a pulse committed to a cable are two: induction, and the electrostatic capacity of the wire, that is, the capacity of the wire to take up a charge of its own, just as if it were the metal of a Leyden jar.

Let us first consider induction. As a current takes its way through the copper core it induces in its surroundings a second and opposing current. For this the remedy is one too costly to be applied. Were a cable manufactured in a double line, as in the best telephonic circuits, induction, with its retarding and quenching effects, would be neutralized. Here the steel wire armour which encircles the cable plays an unwelcome part. Induction is always proportioned to the conductivity of the mass in which it appears; as steel is an excellent conductor, the armour of an ocean cable, close as it is to the copper core, has induced in it a current much stronger, and therefore more retarding, than if the steel wire were absent.

A word now as to the second difficulty in working beneath the sea--that due to the absorbing power of the line itself. An Atlantic cable, like any other extended conductor, is virtually a long, cylindrical Leyden jar, the copper wire forming the inner coat, and its surroundings the outer coat. Before a signal can be received at the distant terminus the wire must first be charged. The effect is somewhat like transmitting a signal through water which fills a rubber tube; first of all the tube is distended, and its compression, or secondary effect, really transmits the impulse. A remedy for this is a condenser formed of alternate sheets of tin-foil and mica, C, connected with the battery, B, so as to balance the electric charge of the cable wire (Fig. 60). In the first Atlantic line an impulse demanded one-seventh of a second for its journey. This was reduced when Mr. Whitehouse made the capital discovery that the speed of a signal is increased threefold when the wire is alternately connected with the zinc and copper poles of the battery. Sir William Thomson ascertained that these successive pulses are most effective when of proportioned lengths. He accordingly devised an automatic transmitter which draws a duly perforated slip of paper under a metallic spring connected with the cable. To-day 250 to 300 letters are sent per minute instead of fifteen, as at first.

In many ways a deep-sea cable exaggerates in an instructive manner the phenomena of telegraphy over long aerial lines. The two ends of a cable may be in regions of widely diverse electrical potential, or pressure, just as the readings of the barometer at these two places may differ much. If a copper wire were allowed to offer itself as a gateless conductor it would equalize these variations of potential with serious injury to itself. Accordingly the rule is adopted of working the cable not directly, as if it were a land line, but indirectly through condensers. As the throb sent through such apparatus is but momentary, the cable is in no risk from the strong currents which would course through it if it were permitted to be an open channel.

[Illustration: Fig. 61.--Reflecting galvanometer L, lamp; N, moving spot of light reflected from mirror]

A serious error in working the first cables was in supposing that they required strong currents as in land lines of considerable length. The very reverse is the fact. Mr. Charles Bright, in Submarine Telegraphs, says:

"Mr. Latimer Clark had the conductor of the 1865 and 1866 lines joined together at the Newfoundland end, thus forming an unbroken length of 3,700 miles in circuit. He then placed some sulphuric acid in a very small silver thimble, with a fragment of zinc weighing a grain or two. By this primitive agency he succeeded in conveying signals through twice the breadth of the

Atlantic Ocean in little more than a second of time after making contact. The deflections were not of a dubious character, but full and strong, from which it was manifest than an even smaller battery would suffice to produce somewhat similar effects."

At first in operating the Atlantic cable a mirror galvanometer was employed as a receiver. The principle of this receiver has often been illustrated by a mischievous boy as, with a slight and almost imperceptible motion of his hand, he has used a bit of looking-glass to dart a ray of reflected sunlight across a wide street or a large room. On the same plan, the extremely minute motion of a galvanometer, as it receives the successive pulsations of a message, is magnified by a weightless lever of light so that the words are easily read by an operator (Fig. 61). This beautiful invention comes from the hands of Sir William Thomson [now Lord Kelvin], who, more than any other electrician, has made ocean telegraphy an established success.

In another receiver, also of his design, the siphon recorder, he began by taking advantage of the fact, observed long before by Bose, that a charge of electricity stimulates the flow of a liquid. In its original form the ink-well into which the siphon dipped was insulated and charged to a high voltage by an influence-machine; the ink, powerfully repelled, was spurted from the siphon point to a moving strip of paper beneath (Fig. 62). It was afterward found better to use a delicate mechanical shaker which throws out the ink in minute drops as the cable current gently sways the siphon back and forth (Fig. 63).

Minute as the current is which suffices for cable telegraphy, it is essential that the metallic circuit be not only unbroken, but unimpaired throughout. No part of his duty has more severely taxed the resources of the electrician than to discover the breaks and leaks in his ocean cables. One of his methods is to pour electricity as it were, into a broken wire, much as if it were a narrow tube, and estimate the length of the wire (and consequently the distance from shore to the defect or break) by the quantity of current required to fill it.

FOOTNOTES:

[1] Henry M. Field, "History of the Atlantic Telegraph." New York: Scribner, 1866.

[2] "A Century of Electricity." Boston, Houghton, Mifflin & Co., 1887.

BELL'S TELEPHONIC RESEARCHES

[From "Bell's Electric Speaking Telephones," by George B. Prescott, copyright by D Appleton & Co., New York, 1884]

In a lecture delivered before the Society of Telegraph Engineers, in London, October 31, 1877, Prof. A. G. Bell gave a history of his researches in telephony, together with the experiments that he was led to undertake in his endeavours to produce a practical system of multiple telegraphy, and to realize also the transmission of articulate speech. After the usual introduction, Professor Bell said in part:

It is to-night my pleasure, as well as duty, to give you some account of the telephonic researches in which I have been so long engaged. Many years ago my attention was directed to the mechanism of speech by my father, Alexander Melville Bell, of Edinburgh, who has made a life-long study of the subject. Many of those present may recollect the invention by my father of a means of representing, in a wonderfully accurate manner, the positions of the vocal organs in forming sounds. Together we carried on quite a number of experiments, seeking to discover the correct mechanism of English and foreign elements of speech, and I remember especially an investigation in which we were engaged concerning the musical relations of vowel sounds. When vocal sounds are whispered, each vowel seems to possess a particular pitch of its own, and by whispering certain vowels in succession a musical scale can be distinctly perceived. Our aim was to determine the natural pitch of each vowel; but unexpected difficulties made their appearance, for many of the vowels seemed to possess a double pitch--one due, probably, to the resonance of the air in the mouth, and the other to the resonance of the air contained in the cavity behind the tongue, comprehending the pharynx and larynx.

I hit upon an expedient for determining the pitch, which, at that time, I thought to be original with myself. It consisted in vibrating a tuning fork in front of the mouth while the positions of the vocal organs for the various vowels were silently taken. It was found that each vowel position caused the reinforcement of some particular fork or forks.

I wrote an account of these researches to Mr. Alex. J. Ellis, of London. In reply, he informed me that the experiments related had already been performed by Helmholtz, and in a much more perfect manner than I had done. Indeed, he said that Helmholtz had not only analyzed the vowel sounds into their constituent musical elements, but had actually performed the synthesis of them.

He had succeeded in producing, artificially, certain of the vowel sounds by causing tuning forks of different pitch to vibrate simultaneously by means of an electric current. Mr. Ellis was kind enough to grant me an interview for the purpose of explaining the apparatus employed by Helmholtz in producing these extraordinary effects, and I spent the greater part of a delightful day with him in investigating the subject. At that time, however, I was too slightly acquainted with the laws of electricity fully to understand the explanations given; but the interview had the effect of arousing my interest in the subjects of sound and electricity, and I did not rest until I had obtained possession of a copy of Helmholtz's great work "The Theory of Tone," and had attempted, in a crude and imperfect manner, it is true, to reproduce his results. While reflecting upon the possibilities of the production of sound by electrical means, it struck me that the principle of vibrating a tuning fork by the intermittent attraction of an electro-magnet might be applied to the electrical production of music.

I imagined to myself a series of tuning forks of different pitches, arranged to vibrate automatically in the manner shown by Helmholtz--each fork interrupting, at every vibration, a voltaic current--and the thought occurred, Why should not the depression of a key like that of a piano direct the interrupted current from any one of these forks, through a telegraph wire, to a series of electro-magnets operating the strings of a piano or other musical instrument, in which case a person might play the tuning fork piano in one place and the music be audible from the electro-magnetic piano in a distant city.

The more I reflected upon this arrangement the more feasible did it seem to me; indeed, I saw no reason why the depression of a number of keys at the tuning fork end of the circuit should not be followed by the audible production of a full chord from the piano in the distant city, each tuning fork affecting at the receiving end that string of the piano with which it was in unison. At this time the interest which I felt in electricity led me to study the various systems of telegraphy in use in this country and in America. I was much struck with the simplicity of the Morse alphabet, and with the fact that it could be read by sound. Instead of having the dots and dashes recorded on paper, the operators were in the habit of observing the duration of the click of the instruments, and in this way were enabled to distinguish by ear the various signals.

It struck me that in a similar manner the duration of a musical note might be made to represent the dot or dash of the telegraph code, so that a person might operate one of the keys of the tuning fork piano referred to above, and the duration of the sound proceeding from the corresponding string of the distant piano be observed by an operator stationed there. It seemed to me that in this way a number of distinct telegraph messages might be sent simultaneously from the tuning fork piano to the other end of the circuit by operators, each manipulating a different key of the instrument. These messages would be read by operators stationed at the distant piano, each receiving operator listening for signals for a certain definite pitch, and ignoring all others. In this way could be accomplished the simultaneous transmission of a number of telegraphic messages along a single wire, the number being limited only by the delicacy of the listener's ear. The idea of increasing the carrying power of a telegraph wire in this way took complete possession of my mind, and it was this practical end that I had in view when I commenced my researches in electric telephony.

In the progress of science it is universally found that complexity leads to simplicity, and in narrating the history of scientific research it is often advisable to begin at the end.

In glancing back over my own researches, I find it necessary to designate, by distinct names, a variety of electrical currents by means of which sounds can be produced, and I shall direct your attention to several distinct species of what may be termed telephonic currents of electricity. In order that the peculiarities of these currents may be clearly understood, I shall project upon the screen a graphical illustration of the different varieties.

The graphical method of representing electrical currents shown in Fig. 1 is the best means I have been able to devise of studying, in an accurate manner, the effects produced by various forms of telephonic apparatus, and it has led me to the conception of that peculiar species of telephonic current, here designated as undulatory, which has rendered feasible the artificial production of articulate speech by electrical means.

A horizontal line (g g') is taken as the zero of current, and impulses of positive electricity are represented above the zero line, and negative impulses below it, or vice versa.

The vertical thickness of any electrical impulse (b or d), measured from the zero line, indicates the intensity of the electrical current at the point observed; and the horizontal extension of the electric line (b or d) indicates the duration of the impulse.

Nine varieties of telephonic currents may be distinguished, but it will only be necessary to show you six of these. The three primary varieties designated as intermittent, pulsatory and undulatory, are represented in lines 1, 2 and 3.

Sub-varieties of these can be distinguished as direct or reversed currents, according as the electrical impulses are all of one kind or are alternately positive and negative. Direct currents may still further be distinguished as positive or negative, according as the impulses are of one kind or of the other.

An intermittent current is characterized by the alternate presence and absence of electricity upon the circuit.

A pulsatory current results from sudden or instantaneous changes in the intensity of a continuous current; and

An undulatory current is a current of electricity, the intensity of which varies in a manner proportional to the velocity of the motion of a particle of air during the production of a sound: thus the curve representing graphically the undulatory current for a simple musical note is the curve expressive of a simple pendulous vibration--that is, a sinusoidal curve.

And here I may remark, that, although the conception of the undulatory current of electricity is entirely original with myself, methods of producing sound by means of intermittent and pulsatory currents have long been known. For instance, it was long since discovered that an electro-magnet gives forth a decided sound when it is suddenly magnetized or demagnetized. When the circuit upon which it is placed is rapidly made and broken, a succession of explosive noises proceeds from the magnet. These sounds produce upon the ear the effect of a musical note when the current is interrupted a sufficient number of times per second....

For several years my attention was almost exclusively directed to the production of an instrument for making and breaking a voltaic circuit with extreme rapidity, to take the place of the transmitting tuning fork used in Helmholtz's researches. Without going into details, I shall merely say that the great defects of this plan of multiple telegraphy were found to consist, first, in the fact that the receiving operators were required to possess a good musical ear in order to discriminate the signals; and secondly, that the signals could only pass in one direction along the line (so that two wires would be necessary in order to complete communication in both directions). The first objection was got over by employing the device which I term a "vibratory circuit breaker," whereby musical signals can be automatically recorded....

I have formerly stated that Helmholtz was enabled to produce vowel sounds artificially by combining musical tones of different pitches and intensities. His apparatus is shown in Fig. 2. Tuning forks of different pitch are placed between the poles of electro-magnets (a1, a2, &c.), and are kept in continuous vibration by the action of an intermittent current from the fork b. Resonators, 1, 2, 3, etc., are arranged so as to reinforce the sounds in a greater or less degree, according as the exterior orifices are enlarged or contracted.

Thus it will be seen that upon Helmholtz's plan the tuning forks themselves produce tones of uniform intensity, the loudness being varied by an external reinforcement; but it struck me that the same results would be obtained, and in a much more perfect manner, by causing the tuning forks themselves to vibrate with different degrees of amplitude. I therefore devised the apparatus shown in Fig. 3, which was my first form of articulating telephone. In this figure a harp of steel rods is employed, attached to the poles of a permanent magnet, N. S. When any one of the rods is thrown into vibration an undulatory current is produced in the coils of the electro-magnet E, and the electro-magnet E' attracts the rods of the harp H' with a varying force, throwing into vibration that rod which is in unison with that vibrating at the other end of the circuit. Not only so, but the amplitude of vibration in the one will determine the amplitude of vibration in the other, for the intensity of the induced current is determined by the amplitude of the inducing vibration, and the amplitude of the vibration at the receiving end depends upon the intensity of the attractive impulses. When we sing into a piano, certain of the strings of the instrument are set in vibration sympathetically by the action of the voice with different degrees of amplitude, and a sound, which is an approximation to the vowel uttered, is produced from the piano. Theory shows that, had the piano a very much larger number of strings to the octave, the vowel sounds would be perfectly reproduced. My idea of the action of the apparatus, shown in Fig. 3, was this: Utter a sound in the neighbourhood of the harp H, and certain of the rods would be thrown into vibration with different amplitudes. At the other end of the circuit the corresponding rods of the harp H would vibrate with their proper relations of force, and the timbre [characteristic quality] of the sound would be reproduced. The expense of constructing such an apparatus as that shown in figure 3 deterred me from making the attempt, and I sought to simplify the apparatus before venturing to have it made.

I have before alluded to the invention by my father of a system of physiological symbols for representing the action of the vocal organs, and I had been invited by the Boston Board of Education to conduct a series of experiments with the system in the Boston school for the deaf and dumb. It is well known that deaf mutes are dumb merely because they are deaf, and that there is no defect in their vocal organs to incapacitate them from utterance. Hence it was thought that my father's system of pictorial symbols, popularly known as visible speech, might prove a means whereby we could teach the deaf and dumb to use their vocal organs and to speak. The great success of these experiments urged upon me the advisability of devising method of exhibiting the vibrations of sound optically, for use in teaching the deaf and dumb. For some time I carried on experiments with the manometric capsule of Konig and with the phonautograph of Len Scott. The scientific apparatus in the Institute of Technology in Boston was freely placed at my disposal for these experiments, and it happened that at that time a student of the Institute of Technology, Mr. Maurey, had invented an improvement upon the phonautograph. He had succeeded in vibrating by the voice a stylus of wood about a foot in length, which was attached to the membrane of the phonautograph, and in this way he had been enabled to obtain enlarged tracings upon a plane surface of smoked glass. With this apparatus I succeeded in producing very beautiful tracings of the vibrations of the air for vowel sounds. Some of these tracings are shown in Fig. 4. I was much struck with this improved form of apparatus, and it occurred to me that there was a remarkable likeness between the manner in which this piece of wood was vibrated by the membrane of the phonautograph and the manner in which the ossiculo [small bones] of the human ear were moved by the tympanic membrane. I determined therefore, to construct a phonautograph modelled still more closely upon the mechanism of the human ear, and for this purpose I sought the assistance of a distinguished aurist in Boston, Dr. Clarence J. Blake. He suggested the use of the human ear itself as a phonautograph, instead of making an artificial imitation of it. The idea was novel and struck me accordingly, and I requested my friend to prepare a specimen for me, which he did. The apparatus, as finally constructed, is shown in Fig. 5. The stapes [inmost of the three auditory ossicles] was removed and a pointed piece of hay about an inch in length was attached to the end of the incus [the middle of the three auditory ossicles]. Upon moistening the membrana tympani [membrane of the ear drum] and the ossicul?with a mixture of glycerine and water the necessary mobility of the parts was obtained, and upon singing into the external artificial ear the piece of hay was thrown into vibration, and tracings were obtained upon a plane surface of smoked glass passed rapidly underneath. While engaged in these experiments I was struck with the remarkable disproportion in weight between the membrane and the bones that were vibrated by it. It occurred to me that if a membrane as thin as tissue paper could control the vibration of bones that were, compared to it, of immense size and weight, why should not a larger and thicker membrane be able to vibrate a piece of iron in front of an electro-magnet, in which case the complication of steel rods shown in my first form of telephone, Fig. 3, could be done away with, and a simple piece of iron attached to a membrane be placed at either end of the telegraphic circuit.

Figure 6 shows the form of apparatus that I was then employing for producing undulatory currents of electricity for the purpose of multiple telegraphy. A steel reed, A, was clamped firmly by one extremity to the uncovered leg h of an electro-magnet E, and the free end of the reed projected above the covered leg. When the reed A was vibrated in any mechanical way the battery current was thrown into waves, and electrical undulations traversed the circuit B E W E', throwing into vibration the corresponding reed A' at the other end of the circuit. I immediately proceeded to put my new idea to the test of practical experiment, and for this purpose I attached the reed A (Fig. 7) loosely by one extremity to the uncovered pole h of the magnet, and fastened the other extremity to the centre of a stretched membrane of goldbeaters' skin n. I presumed that upon speaking in the neighbourhood of the membrane n it would be thrown into vibration and cause the steel reed A to move in a similar manner, occasioning undulations in the electrical current that would correspond to the changes in the density of the air during the production of the sound; and I further thought that the change of the density of the current at the receiving end would cause the magnet there to attract the reed A' in such a manner that it should copy the motion of the reed A, in which case its movements would occasion a sound from the membrane n' similar in timbre to that which had occasioned the original vibration.

The results, however, were unsatisfactory and discouraging. My friend, Mr. Thomas A. Watson, who assisted me in this first experiment, declared that he heard a faint sound proceed from the telephone at his end of the circuit, but I was unable to verify his assertion. After many experiments, attended by the same only partially successful results, I determined to reduce the size and weight of the spring as much as possible. For this purpose I glued a piece of clock spring about the size and shape of my thumb nail, firmly to the centre of the diaphragm, and had a similar instrument at the other end (Fig. 8); we were then enabled to obtain distinctly audible effects. I remember an experiment made with this telephone, which at the time gave me great satisfaction and delight. One of the telephones was placed in my lecture room in the Boston University, and the other in the basement of the adjoining building. One of my students repaired to the distant telephone to observe the effects of articulate speech, while I uttered the sentence, "Do you understand what I say?" into the telephone placed in the lecture hall. To my delight an answer was returned through the instrument itself, articulate sounds proceeded from the steel spring attached to the membrane, and I heard the sentence, "Yes, I understand you perfectly." It is a mistake, however, to suppose that the articulation was by any means perfect, and expectancy no doubt had a great deal to do with my recognition of the sentence; still, the articulation was there, and I recognized the fact that the indistinctness was entirely due to the imperfection of the instrument. I will not trouble you by detailing the various stages through which the apparatus passed, but shall merely say that after a time I produced the form of instrument shown in Fig. 9, which served very well as a receiving telephone. In this condition my invention was, in 1876, exhibited at the Centennial Exhibition in Philadelphia. The telephone shown in Fig. 8 was used as a transmitting instrument, and that in Fig. 9 as a receiver, so that vocal communication was only established in one direction....

The articulation produced from the instrument shown in Fig. 9 was remarkably distinct, but its great defect consisted in the fact that it could not be used as a transmitting instrument, and thus two telephones were required at each station, one for transmitting and one for receiving spoken messages.

It was determined to vary the construction of the telephone shown in Fig. 8, and I sought, by changing the size and tension of the membrane, the diameter and thickness of the steel spring, the size and power of the magnet, and the coils of insulated wire around their poles, to discover empirically the exact effect of each element of the combination, and thus to deduce a more perfect form of apparatus. It was found that a marked increase in the loudness of the sounds resulted from shortening the length of the coils of wire, and by enlarging the iron diaphragm which was glued to the membrane. In the latter case, also, the distinctness of the articulation was improved. Finally, the membrane of goldbeaters' skin was discarded entirely, and a simple iron plate was used instead, and at once intelligible articulation was obtained. The new form of instrument is that shown in Fig. 10, and, as had been long anticipated, it was proved that the only use of the battery was to magnetize the iron core, for the effects were equally audible when the battery was omitted and a rod of magnetized steel substituted for the iron core of the magnet.

It was my original intention, as shown in Fig. 3, and it was always claimed by me, that the final form of telephone would be operated by permanent magnets in place of batteries, and numerous experiments had been carried on by Mr. Watson and myself privately for the purpose of producing this effect.

At the time the instruments were first exhibited in public the results obtained with permanent magnets were not nearly so striking as when a voltaic battery was employed, wherefore we thought it best to exhibit only the latter form of instrument.

The interest excited by the first published accounts of the operation of the telephone led many persons to investigate the subject, and I doubt not that numbers of experimenters have independently discovered that permanent magnets might be employed instead of voltaic batteries. Indeed, one gentleman, Professor Dolbear, of Tufts College, not only claims to have discovered the magneto-electric telephone, but, I understand, charges me with having obtained the idea from him through the medium of a mutual friend.

A still more powerful form of apparatus was constructed by using a powerful compound horseshoe magnet in place of the straight rod which had been previously used (see Fig. 11). Indeed, the sounds produced by means of this instrument were of sufficient loudness to be faintly audible to a large audience, and in this condition the instrument was exhibited in the Essex Institute, in Salem, Massachusetts, on the 12th of February, 1877, on which occasion a short speech shouted into a similar telephone in Boston sixteen miles away, was heard by the audience in Salem. The tones of the speaker's voice were distinctly audible to an audience of six hundred people, but the articulation was only distinct at a distance of about six feet. On the same occasion, also, a report of the lecture was transmitted by word of mouth from Salem to Boston, and published in the papers the next morning.

From the form of telephone shown in Fig. 10 to the present form of the instrument (Fig. 12) is but a step. It is, in fact, the arrangement of Fig. 10 in a portable form, the magnet F. H. being placed inside the handle and a more convenient form of mouthpiece provided....

It was always my belief that a certain ratio would be found between the several parts of a telephone, and that the size of the instrument was immaterial; but Professor Peirce was the first to demonstrate the extreme smallness of the magnets which might be employed. And here, in order to show the parallel lines in which we were working, I may mention the fact that two or three days after I had constructed a telephone of the portable form

(Fig. 12), containing the magnet inside the handle, Dr. Channing was kind enough to send me a pair of telephones of a similar pattern, which had been invented by experimenters at Providence. The convenient form of the mouthpiece shown in Fig. 12, now adopted by me, was invented solely by my friend, Professor Peirce. I must also express my obligations to my friend and associate, Mr. Thomas A. Watson, of Salem, Massachusetts, who has for two years past given me his personal assistance in carrying on my researches.

In pursuing my investigations I have ever had one end in view--the practical improvement of electric telegraphy--but I have come across many facts which, while having no direct bearing upon the subject of telegraphy, may yet possess an interest for you.

For instance, I have found that a musical tone proceeds from a piece of plumbago or retort carbon when an intermittent current of electricity is passed through it, and I have observed the most curious audible effects produced by the passage of reversed intermittent currents through the human body. A breaker was placed in circuit with the primary wires of an induction coil, and the fine wires were connected with two strips of brass. One of these strips was held closely against the ear, and a loud sound proceeded from it whenever the other slip was touched with the other hand. The strips of brass were next held one in each hand. The induced currents occasioned a muscular tremor in the fingers. Upon placing my forefinger to my ear a loud crackling noise was audible, seemingly proceeding from the finger itself. A friend who was present placed my finger to his ear, but heard nothing. I requested him to hold the strips himself. He was then distinctly conscious of a noise (which I was unable to perceive) proceeding from his finger. In this case a portion of the induced current passed through the head of the observer when he placed his ear against his own finger, and it is possible that the sound was occasioned by a vibration of the surfaces of the ear and finger in contact.

When two persons receive a shock from a Ruhmkorff's coil by clasping hands, each taking hold of one wire of the coil with the free hand, a sound proceeds from the clasped hands. The effect is not produced when the hands are moist. When either of the two touches the body of the other a loud sound comes from the parts in contact. When the arm of one is placed against the arm of the other, the noise produced can be heard at a distance of several feet. In all these cases a slight shock is experienced so long as the contact is preserved. The introduction of a piece of paper between the parts in contact does not materially interfere with the production of the sounds, but the unpleasant effects of the shock are avoided.

[Illustration: Fig. 12]

When an intermittent current from a Ruhmkorff's coil is passed through the arms a musical note can be perceived when the ear is closely applied to the arm of the person experimented upon. The sound seems to proceed from the muscles of the fore-arm and from the biceps muscle. Mr. Elisha Gray has also produced audible effects by the passage of electricity through the human body.

An extremely loud musical note is occasioned by the spark of a Ruhmkorff's coil when the primary circuit is made and broken with sufficient rapidity. When two breakers of different pitch are caused simultaneously to open and close the primary circuit a double tone proceeds from the spark.

A curious discovery, which may be of interest to you, has been made by Professor Blake. He constructed a telephone in which a rod of soft iron, about six feet in length, was used instead of a permanent magnet. A friend sang a continuous musical tone into the mouthpiece of a telephone, like that shown in Fig. 12, which was connected with the soft iron instrument alluded to above. It was found that the loudness of the sound produced in this telephone varied with the direction in which the iron rod was held, and that the maximum effect was produced when the rod was in the position of the dipping needle. This curious discovery of Professor Blake has been verified by myself.

When a telephone is placed in circuit with a telegraph line the telephone is found seemingly to emit sounds on its own account. The most extraordinary noises are often produced, the causes of which are at present very obscure. One class of sounds is produced by the inductive influence of neighbouring wires and by leakage from them, the signals of the Morse alphabet passing over neighbouring wires being audible in the telephone, and another class can be traced to earth currents upon the wire, a curious modification of this sound revealing the presence of defective joints in the wire.

Professor Blake informs me that he has been able to use the railroad track for conversational purposes in place of a telegraph wire, and he further states that when only one telephone was connected with the track the sounds of Morse operating were distinctly audible in the telephone, although the nearest telegraph wires were at least fifty feet distant.

Professor Peirce has observed the most singular sounds produced from a telephone in connection with a telegraph wire during the aurora borealis, and I have just heard of a curious phenomenon lately observed by Dr. Channing. In the city of Providence, Rhode Island, there is an over-house wire about one mile in extent with a telephone at either end. On one occasion the sound of music and singing was faintly audible in one of the telephones. It seemed as if some one were practising vocal music with a pianoforte accompaniment. The natural supposition was that experiments were being made with the telephone at the other end of the circuit, but upon inquiry this proved not to have been the case. Attention having thus been directed to the phenomenon, a watch was kept upon the instruments, and upon a subsequent occasion the same fact was observed at both ends of the line by Dr. Channing and his friends. It was proved that the sounds continued for about two hours, and usually commenced about the same time. A searching examination of the line disclosed nothing abnormal in its condition, and I am unable to give you any explanation of this curious phenomenon. Dr. Channing has, however, addressed a letter upon the subject to the editor of one of the Providence papers, giving the names of such songs as were recognized, and full details of the observations, in the hope that publicity may lead to the discovery of the performer, and thus afford a solution of the mystery.

My friend, Mr. Frederick A. Gower, communicated to me a curious observation made by him regarding the slight earth connection required to establish a circuit for the telephone, and together we carried on a series of experiments with rather startling results. We took a couple of telephones and an insulated wire about 100 yards in length into a garden, and were enabled to carry on conversation with the greatest ease when we held in our hands what should have been the earth wire, so that the connection with the ground was formed at either end through our bodies, our feet being clothed with cotton socks and leather boots. The day was fine, and the grass upon which we stood was seemingly perfectly dry. Upon standing upon a gravel walk the vocal sounds, though much diminished, were still perfectly intelligible, and the same result occurred when standing upon a brick wall one foot in height, but no sound was audible when one of us stood upon a block of freestone.

One experiment which we made is so very interesting that I must speak of it in detail. Mr. Gower made earth connection at his end of the line by standing upon a grass plot, whilst at the other end of the line I stood upon a wooden board. I requested Mr. Gower to sing a continuous musical note, and to my surprise the sound was very distinctly audible from the telephone in my hand. Upon examining my feet I discovered that a single blade of grass was bent over the edge of the board, and that my foot touched it. The removal of this blade of grass was followed by the cessation of the sound from the telephone, and I found that the moment I touched with the toe of my boot a blade of grass or the petal of a daisy the sound was again audible.

The question will naturally arise, Through what length of wire can the telephone be used? In reply to this I may say that the maximum amount of resistance through which the undulatory current will pass, and yet retain sufficient force to produce an audible sound at the distant end, has yet to be determined; no difficulty has, however, been experienced in laboratory experiments in conversing through a resistance of 60,000 ohms, which has been the maximum at my disposal. On one occasion, not having a rheostat [for producing resistance] at hand, I passed the current through the bodies of sixteen persons, who stood hand in hand. The longest length of real telegraph line through which I have attempted to converse has been about 250 miles. On this occasion no difficulty was experienced so long as parallel lines were not in operation. Sunday was chosen as the day on which it was probable other circuits would be at rest. Conversation was carried on between myself, in New York, and Mr. Thomas A. Watson, in Boston, until the opening of business upon the other wires. When this happened the vocal sounds were very much diminished, but still audible. It seemed, indeed, like talking through a storm. Conversation, though possible, could be carried on with difficulty, owing to the distracting nature of the interfering currents.

I am informed by my friend Mr. Preece that conversation has been successfully carried on through a submarine cable, sixty miles in length, extending from Dartmouth to the Island of Guernsey, by means of hand telephones.

PHOTOGRAPHING THE UNSEEN: THE ROENTGEN RAY

H. J. W. DAM

In all the history of scientific discovery there has never been, perhaps, so general, rapid, and dramatic an effect wrought on the scientific centres of Europe as has followed, in the past four weeks, upon an announcement made to the Walzburg Physico-Medical Society, at their December [1895] meeting, by Professor William Konrad Roentgen, professor of physics at the Royal University of Walzburg. The first news which reached London was by telegraph from Vienna to the effect that a Professor Roentgen, until then the possessor of only a local fame in the town mentioned, had discovered a new kind of light, which penetrated and photographed through everything. This news was received with a mild interest, some amusement, and much incredulity; and a week passed. Then, by mail and telegraph, came daily clear indications of the stir which the discovery was making in all the great line of universities between Vienna and Berlin. Then Roentgen's own report arrived, so cool, so business-like, and so truly scientific in character, that it left no doubt either of the truth or of the great importance of the preceding reports. To-day, four weeks after the announcement, Roentgen's name is apparently in every scientific publication issued this week in Europe; and accounts of his experiments, of the experiments of others following his method, and of theories as to the strange new force which he has been the first to observe, fill pages of every scientific journal that comes to hand. And before the necessary time elapses for this article to attain publication in America, it is in all ways probable that the laboratories and lecture-rooms of the United States will also be giving full evidence of this contagious arousal of interest over a discovery so strange that its importance cannot yet be measured, its utility be even prophesied, or its ultimate effect upon long established scientific beliefs be even vaguely foretold.

The Roentgen rays are certain invisible rays resembling, in many respects, rays of light, which are set free when a high-pressure electric current is discharged through a vacuum tube. A vacuum tube is a glass tube from which all the air, down to one-millionth of an atmosphere, has been exhausted after the insertion of a platinum wire in either end of the tube for connection with the two poles of a battery or induction coil. When the discharge is sent through the tube, there proceeds from the anode--that is, the wire which is connected with the positive pole of the battery--certain bands of light, varying in colour with the colour of the glass. But these are insignificant in comparison with the brilliant glow which shoots from the cathode, or negative wire. This glow excites brilliant phosphorescence in glass and many substances, and these "cathode rays," as they are called, were observed and studied by Hertz; and more deeply by his assistant, Professor Lenard, Lenard having, in 1894, reported that the cathode rays would penetrate thin films of aluminum, wood, and other substances, and produce photographic results beyond. It was left, however, for Professor Roentgen to discover that during the discharge quite other rays are set free, which differ greatly from those described by Lenard as cathode rays. The most marked difference between the two is the fact that Roentgen rays are not deflected by a magnet, indicating a very essential difference, while their range and penetrative power are incomparably greater. In fact, all those qualities which have lent a sensational character to the discovery of Roentgen's rays were mainly absent from those of Lenard, to the end that, although Roentgen has not been working in an entirely new field, he has by common accord been freely granted all the honors of a great discovery.

Exactly what kind of a force Professor Roentgen has discovered he does not know. As will be seen below, he declines to call it a new kind of light, or a new form of electricity. He has given it the name of the X rays. Others speak of it as the Roentgen rays. Thus far its results only, and not its essence, are known. In the terminology of science it is generally called "a new mode of motion," or, in other words, a new force. As to whether it is or not actually a force new to science, or one of the known forces masquerading under strange conditions, weighty authorities are already arguing. More than one eminent scientist has already affected to see in it a key to the great mystery of the law of gravity. All who have expressed themselves in print have admitted, with more or less frankness, that, in view of Roentgen's discovery, science must forthwith revise, possibly to a revolutionary degree, the long accepted theories concerning the phenomena of light and sound. That the X rays, in their mode of action, combine a strange resemblance to both sound and light vibrations, and are destined to materially affect, if they do not greatly alter, our views of both phenomena, is already certain; and beyond this is the opening into a new and unknown field of physical knowledge, concerning which speculation is already eager, and experimental investigation already in hand, in London, Paris, Berlin, and, perhaps, to a greater or less extent, in every well-equipped physical laboratory in Europe.

This is the present scientific aspect of the discovery. But, unlike most epoch-making results from laboratories, this discovery is one which, to a very unusual degree, is within the grasp of the popular and non-technical imagination. Among the other kinds of matter which these rays penetrate with ease is human flesh. That a new photography has suddenly arisen which can photograph the bones, and, before long, the organs of the human body; that a light has been found which can penetrate, so as to make a photographic record, through everything from a purse or a pocket to the walls of a room or a house, is news which cannot fail to startle everybody. That the eye of the physician or surgeon, long baffled by the skin, and vainly seeking to penetrate the unfortunate darkness of the human body, is now to be supplemented by a camera, making all the parts of the human body as visible, in a way, as the exterior, appears certainly to be a greater blessing to humanity than even the Listerian antiseptic system of surgery; and its benefits must inevitably be greater than those conferred by Lister, great as the latter have been. Already, in the few weeks since Roentgen's announcement, the results of surgical operations under the new system are growing voluminous. In Berlin, not only new bone fractures are being immediately photographed, but joined fractures, as well, in order to examine the results of recent surgical work. In Vienna, imbedded bullets are being photographed, instead of being probed for, and extracted with comparative ease. In London, a wounded sailor, completely paralyzed, whose injury was a mystery, has been saved by the photographing of an object imbedded in the spine, which, upon extraction, proved to be a small knife-blade. Operations for malformations, hitherto obscure, but now clearly revealed by the new photography, are already becoming common, and are being reported from all directions. Professor Czermark of Graz has photographed the living skull, denuded of flesh and hair, and has begun the adaptation of the new photography to brain study. The relation of the new rays to thought rays is being eagerly discussed in what may be called the non-exact circles and journals; and all that numerous group of inquirers into the occult, the believers in clairvoyance, spiritualism, telepathy, and kindred orders of alleged phenomena, are confident of finding in the new force long-sought facts in proof of their claims. Professor Neusser in Vienna has photographed gallstones in the liver of one patient (the stone showing snow-white in the negative), and a stone in the bladder of another patient. His results so far induce him to announce that all the organs of the human body can, and will, shortly, be photographed. Lannelongue of Paris has exhibited to the Academy of Science photographs of bones showing inherited tuberculosis which had not otherwise revealed itself. Berlin has already formed a society of forty for the immediate prosecution of researches into both the character of the new force and its physiological possibilities. In the next few weeks these strange announcements will be trebled or quadrupled, giving the best evidence from all quarters of the great future that awaits the Roentgen rays, and the startling impetus to the universal search for knowledge that has come at the close of the nineteenth century from the modest little laboratory in the Pleicher Ring at Walzburg.

The Physical Institute, Professor Roentgen's particular domain, is a modest building of two stories and basement, the upper story constituting his private residence, and the remainder of the building being given over to lecture rooms, laboratories, and their attendant offices. At the door I was met by an old serving-man of the idolatrous order, whose pain was apparent when I asked for "Professor" Roentgen, and he gently corrected me with "Herr Doctor Roentgen." As it was evident, however, that we referred to the same person, he conducted me along a wide, bare hall, running the length of the building, with blackboards and charts on the walls. At the end he showed me into a small room on the right. This contained a large table desk, and a small table by the window, covered by photographs, while the walls held rows of shelves laden with laboratory and other records. An open door led into a somewhat larger room, perhaps twenty feet by fifteen, and I found myself gazing into a laboratory which was the scene of the discovery--a laboratory which, though in all ways modest, is destined to be enduringly historical.

There was a wide table shelf running along the farther side, in front of the two windows, which were high, and gave plenty of light. In the centre was a stove; on the left, a small cabinet whose shelves held the small objects which the professor had been using. There was a table in the left-hand corner; and another small table--the one on which living bones were first photographed--

was near the stove, and a Ruhmkorff coil was on the right. The lesson of the laboratory was eloquent. Compared, for instance, with the elaborate, expensive, and complete apparatus of, say, the University of London, or of any of the great American universities, it was bare and unassuming to a degree. It mutely said that in the great march of science it is the genius of man, and not the perfection of appliances, that breaks new ground in the great territory of the unknown. It also caused one to wonder at and endeavour to imagine the great things which are to be done through elaborate appliances with the Roentgen rays--a field in which the United States, with its foremost genius in invention, will very possibly, if not probably, take the lead--when the discoverer himself had done so much with so little. Already, in a few weeks, a skilled London operator, Mr. A. A. C. Swinton, has reduced the necessary time of exposure for Roentgen photographs from fifteen minutes to four. He used, however, a Tesla oil coil, discharged by twelve half-gallon Leyden jars, with an alternating current of twenty thousand volts' pressure. Here were no oil coils, Leyden jars, or specially elaborate and expensive machines. There were only a Ruhmkorff coil and Crookes (vacuum) tube and the man himself.

Professor Roentgen entered hurriedly, something like an amiable gust of wind. He is a tall, slender, and loose-limbed man, whose whole appearance bespeaks enthusiasm and energy. He wore a dark blue sack suit, and his long, dark hair stood straight up from his forehead, as if he were permanently electrified by his own enthusiasm. His voice is full and deep, he speaks rapidly, and, altogether, he seems clearly a man who, once upon the track of a mystery which appealed to him, would pursue it with unremitting vigor. His eyes are kind, quick, and penetrating; and there is no doubt that he much prefers gazing at a Crookes tube to beholding a visitor, visitors at present robbing him of much valued time. The meeting was by appointment, however, and his greeting was cordial and hearty. In addition to his own language he speaks French well and English scientifically, which is different from speaking it popularly. These three tongues being more or less within the equipment of his visitor, the conversation proceeded on an international or polyglot basis, so to speak, varying at necessity's demand.

It transpired in the course of inquiry, that the professor is a married man and fifty years of age, though his eyes have the enthusiasm of twenty-five. He was born near Zurich, and educated there, and completed his studies and took his degree at Utrecht. He has been at Walzburg about seven years, and had made no discoveries which he considered of great importance prior to the one under consideration. These details were given under good-natured protest, he failing to understand why his personality should interest the public. He declined to admire himself or his results in any degree, and laughed at the idea of being famous. The professor is too deeply interested in science to waste any time in thinking about himself. His emperor had feasted, flattered, and decorated him, and he was loyally grateful. It was evident, however, that fame and applause had small attractions for him, compared to the mysteries still hidden in the vacuum tubes of the other room.

"Now, then," said he, smiling, and with some impatience, when the preliminary questions at which he chafed were over, "you have come to see the invisible rays."

"Is the invisible visible?"

"Not to the eye; but its results are. Come in here."

From a photograph by A. A. C. Swinton, Victoria Street, London. Exposure, fifty-five seconds]

He led the way to the other square room mentioned, and indicated the induction coil with which his researches were made, an ordinary Ruhmkorff coil, with a spark of from four to six inches, charged by a current of twenty amperes. Two wires led from the coil, through an open door, into a smaller room on the right. In this room was a small table carrying a Crookes tube connected with the coil. The most striking object in the room, however, was a huge and mysterious tin box about seven feet high and four feet square. It stood on end, like a huge packing case, its side being perhaps five inches from the Crookes tube.

The professor explained the mystery of the tin box, to the effect that it was a device of his own for obtaining a portable dark-room. When he began his investigations he used the whole room, as was shown by the heavy blinds and curtains so arranged as to exclude the entrance of all interfering light from the windows. In the side of the tin box, at the point immediately against the tube, was a circular sheet of aluminum one millimetre in thickness, and perhaps eighteen inches in diameter, soldered to the surrounding tin. To study his rays the professor had only to turn on the current, enter the box, close the door, and in perfect darkness inspect only such light or light effects as he had a right to consider his own, hiding his light, in fact, not under the Biblical bushel, but in a more commodious box.

"Step inside," said he, opening the door, which was on the side of the box farthest from the tube. I immediately did so, not altogether certain whether my skeleton was to be photographed for general inspection, or my secret thoughts held up to light on a glass plate. "You will find a sheet of barium paper on the shelf," he added, and then went away to the coil. The door was closed, and the interior of the box became black darkness. The first thing I found was a wooden stool, on which I resolved to sit. Then I found the shelf on the side next the tube, and then the sheet of paper prepared with barium platinocyanide. I was thus being shown the first phenomenon which attracted the discoverer's attention and led to his discovery, namely, the passage of rays, themselves wholly invisible, whose presence was only indicated by the effect they produced on a piece of sensitized photographic paper.

A moment later, the black darkness was penetrated by the rapid snapping sound of the high-pressure current in action, and I knew that the tube outside was glowing. I held the sheet vertically on the shelf, perhaps four inches from the plate. There was no change, however, and nothing was visible.

"Do you see anything?" he called.

"No."

"The tension is not high enough;" and he proceeded to increase the pressure by operating an apparatus of mercury in long vertical tubes acted upon automatically by a weight lever which stood near the coil. In a few moments the sound of the discharge again began, and then I made my first acquaintance with the Roentgen rays.

The moment the current passed, the paper began to glow. A yellowish green light spread all over its surface in clouds, waves and flashes. The yellow-green luminescence, all the stranger and stronger in the darkness, trembled, wavered, and floated over the paper, in rhythm with the snapping of the discharge. Through the metal plate, the paper, myself, and the tin box, the invisible rays were flying, with an effect strange, interesting and uncanny. The metal plate seemed to offer no appreciable resistance to the flying force, and the light was as rich and full as if nothing lay between the paper and the tube.

"Put the book up," said the professor.

I felt upon the shelf, in the darkness, a heavy book, two inches in thickness, and placed this against the plate. It made no difference. The rays flew through the metal and the book as if neither had been there, and the waves of light, rolling cloud-like over the paper, showed no change in brightness. It was a clear, material illustration of the ease with which paper and wood are penetrated. And then I laid book and paper down, and put my eyes against the rays. All was blackness, and I neither saw nor felt anything. The discharge was in full force, and the rays were flying through my head, and, for all I knew, through the side of the box behind me. But they were invisible and impalpable. They gave no sensation whatever. Whatever the mysterious rays may be, they are not to be seen, and are to be judged only by their works.

I was loath to leave this historical tin box, but time pressed. I thanked the professor, who was happy in the reality of his discovery and the music of his sparks. Then I said: "Where did you first photograph living bones?"

"Here," he said, leading the way into the room where the coil stood. He pointed to a table on which was another--the latter a small short-legged wooden one with more the shape and size of a wooden seat. It was two feet square and painted coal black. I viewed it with interest. I would have bought it, for the little table on which light was first sent through the human body will some day be a great historical curiosity; but it was not for sale. A photograph of it would have been a consolation, but for several reasons one was not to be had at present. However, the historical table was there, and was duly inspected.

"How did you take the first hand photograph?" I asked.

The professor went over to a shelf by the window, where lay a number of prepared glass plates, closely wrapped in black paper. He put a Crookes tube underneath the table, a few inches from the under side of its top. Then he laid his hand flat on the top of the table, and placed the glass plate loosely on his hand.

"You ought to have your portrait painted in that attitude," I suggested.

"No, that is nonsense," said he, smiling.

"Or be photographed." This suggestion was made with a deeply hidden purpose.

The rays from the Roentgen eyes instantly penetrated the deeply hidden purpose. "Oh, no," said he; "I can't let you make pictures of me. I am too busy." Clearly the professor was entirely too modest to gratify the wishes of the curious world.

"Now, Professor," said I, "will you tell me the history of the discovery?"

"There is no history," he said. "I have been for a long time interested in the problem of the cathode rays from a vacuum tube as studied by Hertz and Lenard. I had followed their and other researches with great interest, and determined, as soon as I had the time, to make some researches of my own. This time I found at the close of last October. I had been at work for some days when I discovered something new."

"What was the date?"

"The eighth of November."

"And what was the discovery?"

"I was working with a Crookes tube covered by a shield of black cardboard. A piece of barium platinocyanide paper lay on the bench there. I had been passing a current through the tube, and I noticed a peculiar black line across the paper."

"What of that?"

"The effect was one which could only be produced, in ordinary parlance, by the passage of light. No light could come from the tube, because the shield which covered it was impervious to any light known, even that of the electric arc."

"And what did you think?"

"I did not think; I investigated. I assumed that the effect must have come from the tube, since its character indicated that it could come from nowhere else. I tested it. In a few minutes there was no doubt about it. Rays were coming from the tube which had a luminescent effect upon the paper. I tried it successfully at greater and greater distances, even at two metres. It seemed at first a new kind of invisible light. It was clearly something new, something unrecorded."

"Is it light?"

"No."

"Is it electricity?"

"Not in any known form."

"What is it?"

"I don't know."

And the discoverer of the X rays thus stated as calmly his ignorance of their essence as has everybody else who has written on the phenomena thus far.

"Having discovered the existence of a new kind of rays, I of course began to investigate what they would do." He took up a series of cabinet-sized photographs. "It soon appeared from tests that the rays had penetrative powers to a degree hitherto unknown. They penetrated paper, wood, and cloth with ease; and the thickness of the substance made no perceptible difference, within reasonable limits." He showed photographs of a box of laboratory weights of platinum, aluminum, and brass, they and the brass hinges all having been photographed from a closed box, without any indication of the box. Also a photograph of a coil of fine wire, wound on a wooden spool, the wire having been photographed, and the wood omitted. "The rays," he continued, "passed through all the metals tested, with a facility varying, roughly speaking, with the density of the metal. These phenomena I have discussed carefully in my report to the Walzburg society, and you will find all the technical results therein stated." He showed a photograph of a small sheet of zinc. This was composed of smaller plates soldered laterally with solders of different metallic proportions. The differing lines of shadow, caused by the difference in the solders, were visible evidence that a new means of detecting flaws and chemical variations in metals had been found. A photograph of a compass showed the needle and dial taken through the closed brass cover. The markings of the dial were in red metallic paint, and thus interfered with the rays, and were reproduced. "Since the rays had this great penetrative power, it seemed natural that they should penetrate flesh, and so it proved in photographing the hand, as I showed you."

A detailed discussion of the characteristics of his rays the professor considered unprofitable and unnecessary. He believes, though, that these mysterious radiations are not light, because their behaviour is essentially different from that of light rays, even those light rays which are themselves invisible. The Roentgen rays cannot be reflected by reflecting surfaces, concentrated by lenses, or refracted or diffracted. They produce photographic action on a sensitive film, but their action is weak as yet, and herein lies the first important field of their development. The professor's exposures were comparatively long--an average of fifteen minutes in easily penetrable media, and half an hour or more in photographing the bones of the hand. Concerning vacuum tubes, he said that he preferred the Hittorf, because it had the most perfect vacuum, the highest degree of air exhaustion being the consummation most desirable. In answer to a question, "What of the future?" he said:

"I am not a prophet, and I am opposed to prophesying. I am pursuing my investigations, and as fast as my results are verified I shall make them public."

"Do you think the rays can be so modified as to photograph the organs of the human body?"

In answer he took up the photograph of the box of weights. "Here are already modifications," he said, indicating the various degrees of shadow produced by the aluminum, platinum, and brass weights, the brass hinges, and even the metallic stamped lettering on the cover of the box, which was faintly perceptible.

"But Professor Neusser has already announced that the photographing of the various organs is possible."

"We shall see what we shall see," he said. "We have the start now; the development will follow in time."

"You know the apparatus for introducing the electric light into the stomach?"

"Yes."

"Do you think that this electric light will become a vacuum tube for photographing, from the stomach, any part of the abdomen or thorax?"

The idea of swallowing a Crookes tube, and sending a high frequency current down into one's stomach, seemed to him exceedingly funny. "When I have done it, I will tell you," he said, smiling, resolute in abiding by results.

"There is much to do, and I am busy, very busy," he said in conclusion. He extended his hand in farewell, his eyes already wandering toward his work in the inside room. And his visitor promptly left him; the words, "I am busy," said in all sincerity, seeming to describe in a single phrase the essence of his character and the watchword of a very unusual man.

Returning by way of Berlin, I called upon Herr Spies of the Urania, whose photographs after the Roentgen method were the first made public, and have been the best seen thus far. In speaking of the discovery he said:

"I applied it, as soon as the penetration of flesh was apparent, to the photograph of a man's hand. Something in it had pained him for years, and the photograph at once exhibited a small foreign object, as you can see;" and he exhibited a copy of the photograph in question. "The speck there is a small piece of glass, which was immediately extracted, and which, in all probability, would have otherwise remained in the man's hand to the end of his days." All of which indicates that the needle which has pursued its travels in so many persons, through so many years, will be suppressed by the camera.

"My next object is to photograph the bones of the entire leg," continued Herr Spies. "I anticipate no difficulty, though it requires some thought in manipulation."

It will be seen that the Roentgen rays and their marvellous practical possibilities are still in their infancy. The first successful modification of the action of the rays so that the varying densities of bodily organs will enable them to be photographed will bring all such morbid growths as tumours and cancers into the photographic field, to say nothing of vital organs which may be abnormally developed or degenerate. How much this means to medical and surgical practice it requires little imagination to conceive. Diagnosis, long a painfully uncertain science, has received an unexpected and wonderful assistant; and how greatly the world will benefit thereby, how much pain will be saved, only the future can determine. In science a new door has been opened where none was known to exist, and a side-light on phenomena has appeared, of which the results may prove as penetrating and astonishing as the Roentgen rays themselves. The most agreeable feature of the discovery is the opportunity it gives for other hands to help; and the work of these hands will add many new words to the dictionaries, many new facts to science, and, in the years long ahead of us, fill many more volumes than there are paragraphs in this brief and imperfect account.

THE WIRELESS TELEGRAPH

GEORGE ILES

[From "Flame, Electricity and the Camera," copyright by Doubleday, Page & Co., New York.]

In a series of experiments interesting enough but barren of utility, the water of a canal, river, or bay has often served as a conductor for the telegraph. Among the electricians who have thus impressed water into their service was Professor Morse. In 1842 he sent a few signals across the channel from Castle Garden, New York, to Governor's Island, a distance of a mile. With much better results, he sent messages, later in the same year, from one side of the canal at Washington to the other, a distance of eighty feet, employing large copper plates at each terminal. The enormous current required to overcome the resistance of water has barred this method from practical adoption.

We pass, therefore, to electrical communication as effected by induction-- the influence which one conductor exerts on another through an intervening insulator. At the outset we shall do well to bear in mind that magnetic phenomena, which are so closely akin to electrical, are always inductive. To observe a common example of magnetic induction, we have only to move a horseshoe magnet in the vicinity of a compass needle, which will instantly sway about as if blown hither and thither by a sharp draught of air. This action takes place if a slate, a pane of glass, or a shingle is interposed between the needle and its perturber. There is no known insulator for magnetism, and an induction of this kind exerts itself perceptibly for many yards when large masses of iron are polarised, so that the derangement of compasses at sea from moving iron objects aboard ship, or from ferric ores underlying a sea-coast, is a constant peril to the mariner.

Electrical conductors behave much like magnetic masses. A current conveyed by a conductor induces a counter-current in all surrounding bodies, and in a degree proportioned to their conductive power. This effect is, of course, greatest upon the bodies nearest at hand, and we have already remarked its serious retarding effect in ocean telegraphy. When the original current is of high intensity, it can induce a perceptible current in another wire at a distance of several miles. In 1842 Henry remarked that electric waves had this quality, but in that early day of electrical interpretation the full significance of the fact eluded him. In the top room of his house he produced a spark an inch long, which induced currents in wires stretched in his cellar, through two thick floors and two rooms which came between. Induction of this sort causes the annoyance, familiar in single telephonic circuits, of being obliged to overhear other subscribers, whose wires are often far away from our own.

The first practical use of induced currents in telegraphy was when Mr. Edison, in 1885, enabled the trains on a line of the Staten Island Railroad to be kept in constant communication with a telegraphic wire, suspended in the ordinary way beside the track. The roof of a car was of insulated metal, and every tap of an operator's key within the walls electrified the roof just long enough to induce a brief pulse through the telegraphic circuit. In sending a message to the car this wire was, moment by moment, electrified, inducing a response first in the car roof, and next in the "sounder" beneath it. This remarkable apparatus, afterward used on the Lehigh Valley Railroad, was discontinued from lack of commercial support, although it would seem to be advantageous to maintain such a service on other than commercial grounds. In case of chance obstructions on the track, or other peril, to be able to communicate at any moment with a train as it speeds along might mean safety instead of disaster. The chief item in the cost of this system is the large outlay for a special telegraphic wire.

The next electrician to employ induced currents in telegraphy was Mr. (now Sir) William H. Preece, the engineer then at the head of the British telegraph system. Let one example of his work be cited. In 1896 a cable was laid between Lavernock, near Cardiff, on the Bristol Channel, and Flat Holme, an island three and a third miles off. As the channel at this point is a much-frequented route and anchor ground, the cable was broken again and again. As a substitute for it Mr. Preece, in 1898, strung wires along the opposite shores, and found that an electric pulse sent through one wire instantly made itself heard in a telephone connected with the other. It would seem that in this etheric form of telegraphy the two opposite lines of wire must be each as long as the distance which separates them; therefore, to communicate across the English Channel from Dover to Calais would require a line along each coast at least twenty miles in length. Where such lines exist for ordinary telegraphy, they might easily lend themselves to the Preece system of signalling in case a submarine cable were to part.

Marconi, adopting electrostatic instead of electro-magnetic waves, has won striking results. Let us note the chief of his forerunners, as they prepared the way for him. In 1864 Maxwell observed that electricity and light have the same velocity, 186,400 miles a second, and he formulated the theory that electricity propagates itself in waves which differ from those of light only in being longer. This was proved to be true by Hertz, who in 1888 showed that where alternating currents of very high frequency were set up in an open circuit, the energy might be conveyed entirely away from the circuit into the surrounding space as electric waves. His detector was a nearly closed circle of wire, the ends being soldered to metal balls almost in contact. With this simple apparatus he demonstrated that electric waves move with the speed of light, and that they can be reflected and refracted precisely as if they formed a visible beam. At a certain intensity of strain the air insulation broke down, and the air became a conductor. This phenomenon of passing quite suddenly from a non-conductive to a conductive state is, as we shall duly see, also to be noted when air or other gases are exposed to the X ray.

Now for the effect of electric waves such as Hertz produced, when they impinge upon substances reduced to powder or filings. Conductors, such as the metals, are of inestimable service to the electrician; of equal value are non-conductors, such as glass and gutta-percha, as they strictly fence in an electric stream. A third and remarkable vista opens to experiment when it deals with substances which, in their normal state, are non-conductive, but which, agitated by an electric wave, instantly become conductive in a high degree. As long ago as 1866 Mr. S. A. Varley noticed that black lead, reduced to a loose dust, effectually intercepted a current from fifty Daniell cells, although the battery poles were very near each other. When he increased the electric tension four- to six-fold, the black-lead particles at once compacted themselves so as to form a bridge of excellent conductivity. On this principle he invented a lightning-protector for electrical instruments, the incoming flash causing a tiny heap of carbon dust to provide it with a path through which it could safely pass to the earth. Professor Temistocle Calzecchi Onesti of Fermo, in 1885, in an independent series of researches, discovered that a mass of powdered copper is a non-conductor until an electric wave beats upon it; then, in an instant, the mass resolves itself into a conductor almost as efficient as if it were a stout, unbroken wire. Professor Edouard Branly of Paris, in 1891, on this principle devised a coherer, which passed from resistance to invitation when subjected to an electric impulse from afar. He enhanced the value of his device by the vital discovery that the conductivity bestowed upon filings by electric discharges could be destroyed by simply shaking or tapping them apart.

In a homely way the principle of the coherer is often illustrated in ordinary telegraphic practice. An operator notices that his instrument is not working well, and he suspects that at some point in his circuit there is a defective contact. A little dirt, or oxide, or dampness, has come in between two metallic surfaces; to be sure, they still touch each other, but not in the firm and perfect way demanded for his work. Accordingly he sends a powerful current abruptly into the line, which clears its path thoroughly, brushes aside dirt, oxide, or moisture, and the circuit once more is as it should be. In all likelihood, the coherer is acted upon in the same way. Among the physicists who studied it in its original form was Dr. Oliver J. Lodge. He improved it so much that, in 1894, at the Royal Institution in London, he was able to show it as an electric eye that registered the impact of invisible rays at a distance of more than forty yards. He made bold to say that this distance might be raised to half a mile.

As early as 1879 Professor D. E. Hughes began a series of experiments in wireless telegraphy, on much the lines which in other hands have now reached commercial as well as scientific success. Professor Hughes was the inventor of the microphone, and that instrument, he declared, affords an unrivalled means of receiving wireless messages, since it requires no tapping to restore its non-conductivity. In his researches this investigator was convinced that his signals were propagated, not by electro-magnetic induction, but by aerial electric waves spreading out from an electric spark. Early in 1880 he showed his apparatus to Professor Stokes, who observed its operation carefully. His dictum was that he saw nothing which could not be explained by known electro-magnetic effects. This erroneous judgment so discouraged Professor Hughes that he desisted from following up his experiments, and thus, in all probability, the birth of the wireless telegraph was for several years delayed.[3]

The coherer, as improved by Marconi, is a glass tube about one and one-half inches long and about one-twelfth of an inch in internal diameter. The electrodes are inserted in this tube so as almost to touch; between them is about one-thirtieth of an inch filled with a pinch of the responsive mixture which forms the pivot of the whole contrivance. This mixture is 90 per cent. nickel filings, 10 per cent. hard silver filings, and a mere trace of mercury; the tube is exhausted of air to within one ten-thousandth part (Fig. 71). How does this trifle of metallic dust manage loudly to utter its signals through a telegraphic sounder, or forcibly indent them upon a moving strip of paper? Not directly, but indirectly, as the very last refinement of initiation. Let us imagine an ordinary telegraphic battery strong enough loudly to tick out a message. Be it ever so strong it remains silent until its circuit is completed, and for that completion the merest touch suffices. Now the thread of dust in the coherer forms part of such a telegraphic circuit: as loose dust it is an effectual bar and obstacle, under the influence of electric waves from afar it changes instantly to a coherent metallic link which at once completes the circuit and delivers the message.

An electric impulse, almost too attenuated for computation, is here able to effect such a change in a pinch of dust that it becomes a free avenue instead of a barricade. Through that avenue a powerful blow from a local store of energy makes itself heard and felt. No device of the trigger class is comparable with this in delicacy. An instant after a signal has taken its way through the coherer a small hammer strikes the tiny tube, jarring its particles asunder, so that they resume their normal state of high resistance. We may well be astonished at the sensitiveness of the metallic filings to an electric wave originating many miles away, but let us remember how clearly the eye can see a bright lamp at the same distance as it sheds a sister beam. Thus far no substance has been discovered with a mechanical responsiveness to so feeble a ray of light; in the world of nature and art the coherer stands alone. The electric waves employed by Marconi are about four feet long, or have a frequency of about 250,000,000 per second. Such undulations pass readily through brick or stone walls, through common roofs and floors--indeed, through all substances which are non-conductive to electric waves of ordinary length. Were the energy of a Marconi sending-instrument applied to an arc-lamp, it would generate a beam of a thousand candle-power. We have thus a means of comparing the sensitiveness of the retina to light with the responsiveness of the Marconi coherer to electric waves, after both radiations have undergone a journey of miles.

An essential feature of this method of ethcric telegraphy, due to Marconi himself, is the suspension of a perpendicular wire at each terminus, its length twenty feet for stations a mile apart, forty feet for four miles, and so on, the telegraphic distance increasing as the square of the length of suspended wire. In the Kingstown regatta, July, 1898, Marconi sent from a yacht under full steam a report to the shore without the loss of a moment from start to finish. This feat was repeated during the protracted contest between the Columbia and the Shamrock yachts in New York Bay, October, 1899. On March 28, 1899, Marconi signals put Wimereux, two miles north of Boulogne, in communication with the South Foreland Lighthouse, thirty-two miles off.[4] In August, 1899, during the manoeuvres of the British navy, similar messages were sent as far as eighty miles. It was clearly demonstrated that a new power had been placed in the hands of a naval commander. "A touch on a button in a flagship is all that is now needed to initiate every tactical evolution in a fleet, and insure an almost automatic precision in the resulting movements of the ships. The flashing lantern is superseded at night, flags and the semaphore by day, or, if these are retained, it is for services purely auxiliary. The hideous and bewildering shrieks of the steam-siren need no longer be heard in a fog, and the uncertain system of gun signals will soon become a thing of the past." The interest of the naval and military strategist in the Marconi apparatus extends far beyond its communication of intelligence. Any electrical appliance whatever may be set in motion by the same wave that actuates a telegraphic sounder. A fuse may be ignited, or a motor started and directed, by apparatus connected with the coherer, for all its minuteness. Mr. Walter Jamieson and Mr. John Trotter have devised means for the direction of torpedoes by ether waves, such as those set at work in the wireless telegraph. Two rods projecting above the surface of the water receive the waves, and are in circuit with a coherer and a relay. At the will of the distant operator a hollow wire coil bearing a current draws in an iron core either to the right or to the left, moving the helm accordingly.

As the news of the success of the Marconi telegraph made its way to the London Stock Exchange there was a fall in the shares of cable companies. The fear of rivalry from the new invention was baseless. As but fifteen words a minute are transmissible by the Marconi system, it evidently does not compete with a cable, such as that between France and England, which can transmit 2,500 words a minute without difficulty. The Marconi telegraph comes less as a competitor to old systems than as a mode of communication which creates a field of its own. We have seen what it may accomplish in war, far outdoing any feat possible to other apparatus, acoustic, luminous, or electrical. In quite as striking fashion does it break new ground in the service of commerce and trade. It enables lighthouses continually to spell their names, so that receivers aboard ship may give the steersmen their bearings even in storm and fog. In the crowded condition of the steamship "lanes" which cross the Atlantic, a priceless security against collision is afforded the man at the helm. On November 15, 1899, Marconi telegraphed from the American liner St. Paul to the Needles, sixty-six nautical miles away. On December 11 and 12, 1901, he received wireless signals near St. John's, Newfoundland, sent from Poldhu, Cornwall, England, or a distance of 1,800 miles,--a feat which astonished the world. In many cases the telegraphic business to an island is too small to warrant the laying of a cable; hence we find that Trinidad and Tobago are to be joined by the wireless system, as also five islands of the Hawaiian group, eight to sixty-one miles apart.

A weak point in the first Marconi apparatus was that anybody within the working radius of the sending-instrument could read its messages. To modify this objection secret codes were at times employed, as in commerce and diplomacy. A complete deliverance from this difficulty is promised in attuning a transmitter and a receiver to the same note, so that one receiver, and no other, shall respond to a particular frequency of impulses. The experiments which indicate success in this vital particular have been conducted by Professor Lodge.

When electricians, twenty years ago, committed energy to a wire and thus enabled it to go round a corner, they felt that they had done well. The Hertz waves sent abroad by Marconi ask no wire, as they find their way, not round a corner, but through a corner. On May 1, 1899, a party of French officers on board the Ibis at Sangatte, near Calais, spoke to Wimereux by means of a Marconi apparatus, with Cape Grisnez, a lofty promontory, intervening. In ascertaining how much the earth and the sea may obstruct the waves of Hertz there is a broad and fruitful field for investigation. "It may be," says Professor John Trowbridge, "that such long electrical waves roll around the surface of such obstructions very much as waves of sound and of water would do."

It is singular how discoveries sometimes arrive abreast of each other so as to render mutual aid, or supply a pressing want almost as soon as it is felt. The coherer in its present form is actuated by waves of comparatively low frequency, which rise from zero to full height in extremely brief periods, and are separated by periods decidedly longer (Fig. 73). What is needed is a plan by which the waves may flow either continuously or so near together that they may lend themselves to attuning. Dr. Wehnelt, by an extraordinary discovery, may, in all likelihood, provide the lacking device in the form of his interrupter, which breaks an electric circuit as often as two thousand times a second. The means for this amazing performance are simplicity itself (Fig. 74). A jar, a, containing a solution of sulphuric acid has two electrodes immersed in it; one of them is a lead plate of large surface, b; the other is a small platinum wire which protrudes from a glass tube, d. A current passing through the cell between the two metals at c is interrupted, in ordinary cases five hundred times a second, and in extreme cases four times as often, by bubbles of gas given off from the wire instant by instant.

FOOTNOTES:

[3] "History of the Wireless Telegraph," by J. J. Fahie. Edinburgh and London, William Blackwood & Sons; New York, Dodd, Mead & Co., 1899. This work is full of interesting detail, well illustrated.

[4] The value of wireless telegraphy in relation to disasters at sea was proved in a remarkable way yesterday morning. While the Channel was enveloped in a dense fog, which had lasted throughout the greater part of the night, the East Goodwin Lightship had a very narrow escape from sinking at her moorings by being run into by the steamship R. F. Matthews, 1,964 tons gross burden, of London, outward bound from the Thames. The East Goodwin Lightship is one of four such vessels marking the Goodwin Sands, and, curiously enough, it happens to be the one ship which has been fitted out with Signor Marconi's installation for wireless telegraphy. The vessel was moored about twelve miles to the northeast of the South Foreland Lighthouse (where there is another wireless-telegraphy installation), and she is about ten miles from the shore, being directly opposite Deal. The information regarding the collision was at once communicated by wireless telegraphy from the disabled lightship to the South Foreland Lighthouse, where Mr. Bullock, assistant to Signor Marconi, received the following message: "We have just been run into by the steamer R. F. Matthews of London. Steamship is standing by us. Our bows very badly damaged." Mr. Bullock immediately forwarded this information to the Trinity House authorities at Ramsgate.--Times, April 29, 1899.

ELECTRICITY, WHAT ITS MASTERY MEANS: WITH A REVIEW AND A PROSPECT

GEORGE ILES

[From "Flame, Electricity and the Camera," copyright by Doubleday, Page & Co., New York.]

With the mastery of electricity man enters upon his first real sovereignty of nature. As we hear the whirr of the dynamo or listen at the telephone, as we turn the button of an incandescent lamp or travel in an electromobile, we are partakers in a revolution more swift and profound than has ever before been enacted upon earth. Until the nineteenth century fire was justly accounted the most useful and versatile servant of man. To-day electricity is doing all that fire ever did, and doing it better, while it accomplishes uncounted tasks far beyond the reach of flame, however ingeniously applied. We may thus observe under our eyes just such an impetus to human intelligence and power as when fire was first subdued to the purposes of man, with the immense advantage that, whereas the subjugation of fire demanded ages of weary and uncertain experiment, the mastery of electricity is, for the most part, the assured work of the nineteenth century, and, in truth, very largely of its last three decades. The triumphs of the electrician are of absorbing interest in themselves, they bear a higher significance to the student of man as a creature who has gradually come to be what he is. In tracing the new horizons won by electric science and art, a beam of light falls on the long and tortuous paths by which man rose to his supremacy long before the drama of human life had been chronicled or sung.

Of the strides taken by humanity on its way to the summit of terrestrial life, there are but four worthy of mention as preparing the way for the victories of the electrician--the attainment of the upright attitude, the intentional kindling of fire, the maturing of emotional cries to articulate speech, and the invention of written symbols for speech. As we examine electricity in its fruitage we shall find that it bears the unfailing mark of every other decisive factor of human advance: its mastery is no mere addition to the resources of the race, but a multiplier of them. The case is not as when an explorer discovers a plant hitherto unknown, such as Indian corn, which takes its place beside rice and wheat as a new food, and so measures a service which ends there. Nor is it as when a prospector comes upon a new metal, such as nickel, with the sole effect of increasing the variety of materials from which a smith may fashion a hammer or a blade. Almost infinitely higher is the benefit wrought when energy in its most useful phase is, for the first time, subjected to the will of man, with dawning knowledge of its unapproachable powers. It begins at once to marry the resources of the mechanic and the chemist, the engineer and the artist, with issue attested by all its own fertility, while its rays reveal province after province undreamed of, and indeed unexisting, before its advent.

Every other primal gift of man rises to a new height at the bidding of the electrician. All the deftness and skill that have followed from the upright attitude, in its creation of the human hand, have been brought to a new edge and a broader range through electric art. Between the uses of flame and electricity have sprung up alliances which have created new wealth for the miner and the metal-worker, the manufacturer and the shipmaster, with new insights for the man of research. Articulate speech borne on electric waves makes itself heard half-way across America, and words reduced to the symbols of symbols--expressed in the perforations of a strip of paper--take flight through a telegraph wire at twenty-fold the pace of speech. Because the latest leap in knowledge and faculty has been won by the electrician, he has widened the scientific outlook vastly more than any explorer who went before. Beyond any predecessor, he began with a better equipment and a larger capital to prove the gainfulness which ever attends the exploiting a supreme agent of discovery.

As we trace a few of the unending interlacements of electrical science and art with other sciences and arts, and study their mutually stimulating effects, we shall be reminded of a series of permutations where the latest of the factors, because latest, multiplies all prior factors in an unexampled degree.[5] We shall find reason to believe that this is not merely a suggestive analogy, but really true as a tendency, not only with regard to man's gains by the conquest of electricity, but also with respect to every other signal victory which has brought him to his present pinnacle of discernment and rule. If this permutative principle in former advances lay undetected, it stands forth clearly in that latest accession to skill and interpretation which has been ushered in by Franklin and Volta, Faraday and Henry.

Although of much less moment than the triumphs of the electrician, the discovery of photography ranks second in importance among the scientific feats of the nineteenth century. The camera is an artificial eye with almost every power of the human retina, and with many that are denied to vision-- however ingeniously fortified by the lens-maker. A brief outline of photographic history will show a parallel to the permutative impulse so conspicuous in the progress of electricity. At the points where the electrician and the photographer collaborate we shall note achievements such as only the loftiest primal powers may evoke.

A brief story of what electricity and its necessary precursor, fire, have done and promise to do for civilization, may have attraction in itself; so, also, may a review, though most cursory, of the work of the camera and all that led up to it: for the provinces here are as wide as art and science, and their bounds comprehend well-nigh the entirety of human exploits. And between the lines of this story we may read another--one which may tell us something of the earliest stumblings in the dawn of human faculty. When we compare man and his next of kin, we find between the two a great gulf, surely the widest betwixt any allied families in nature. Can a being of intellect, conscience, and aspiration have sprung at any time, however remote, from the same stock as the orang and the chimpanzee? Since 1859, when Darwin published his "Origin of Species," the theory of evolution has become so generally accepted that to-day it is little more assailed than the doctrine of gravitation. And yet, while the average man of intelligence bows to the formula that all which now exists has come from the simplest conceivable state of things,--a universal nebula, if you will,--in his secret soul he makes one exception--himself. That there is a great deal more assent than conviction in the world is a chiding which may come as justly from the teacher's table as from the preacher's pulpit. Now, if we but catch the meaning of man's mastery of electricity, we shall have light upon his earlier steps as a fire-kindler, and as a graver of pictures and symbols on bone and rock. As we thus recede from civilization to primeval savagery, the process of the making of man may become so clear that the arguments of Darwin shall be received with conviction, and not with silent repulse.

As we proceed to recall, one by one, the salient chapters in the history of fire, and of the arts of depiction that foreran the camera, we shall perceive a truth of high significance. We shall see that, while every new faculty has its roots deep in older powers, and while its growth may have been going on for age after age, yet its flowering may be as the event of a morning. Even as our gardens show us the century-plants, once supposed to bloom only at the end of a hundred years, so history, in the large, exhibits discoveries whose harvests are gathered only after the lapse of eons instead of years. The arts of fire were slowly elaborated until man had produced the crucible and the still, through which his labours culminated in metals purified, in acids vastly more corrosive than those of vegetation, in glass and porcelain equally resistant to flame and the electric wave. These were combined in an hour by Volta to build his cell, and in that hour began a new era for human faculty and insight.

It is commonly imagined that the progress of humanity has been at a tolerably uniform pace. Our review of that progress will show that here and there in its course have been leaps, as radically new forces have been brought under the dominion of man. We of the electric revolution are sharply marked off from our great-grandfathers, who looked upon the cell of Volta as a curious toy. They, in their turn, were profoundly differenced from the men of the seventeenth century, who had not learned that flame could outvie the horse as a carrier, and grind wheat better than the mill urged by the breeze. And nothing short of an abyss stretches between these men and their remote ancestors, who had not found a way to warm their frosted fingers or lengthen with lamp or candle the short, dark days of winter.

Throughout the pages of this book there will be some recital of the victories won by the fire-maker, the electrician, the photographer, and many more in the peerage of experiment and research. Underlying the sketch will appear the significant contrast betwixt accessions of minor and of supreme dignity. The finding a new wood, such as that of the yew, means better bows for the archer, stronger handles for the tool-maker; the subjugation of a universal force such as fire, or electricity, stands for the exaltation of power in every field of toil, for the creation of a new earth for the worker, new heavens for the thinker. As a corollary, we shall observe that an increasing width of gap marks off the successive stages of human progress from each other, so that its latest stride is much the longest and most decisive. And it will be further evident that, while every new faculty is of age-long derivation from older powers and ancient aptitudes, it nevertheless comes to the birth in a moment, as it were, and puts a strain of probably fatal severity on those contestants who miss the new gift by however little. We shall, therefore, find that the principle of permutation, here merely indicated, accounts in large measure for three cardinal facts in the history of man: First, his leaps forward; second, the constant accelerations in these leaps; and third, the gap in the record of the tribes which, in the illimitable past, have succumbed as forces of a new edge and sweep have become engaged in the fray.[6]

The interlacements of the arts of fire and of electricity are intimate and pervasive. While many of the uses of flame date back to the dawn of human skill, many more have become of new and higher value within the last hundred years. Fire to-day yields motive power with tenfold the economy of a hundred years ago, and motive power thus derived is the main source of modern electric currents. In metallurgy there has long been an unwitting preparation for the advent of the electrician, and here the services of fire within the nineteenth century have won triumphs upon which the later successes of electricity largely proceed. In producing alloys, and in the singular use of heat to effect its own banishment, novel and radical developments have been recorded within the past decade or two. These, also, make easier and bolder the electrician's tasks. The opening chapters of this book will, therefore, bestow a glance at the principal uses of fire as these have been revealed and applied. This glance will make clear how fire and electricity supplement each other with new and remarkable gains, while in other fields, not less important, electricity is nothing else than a supplanter of the very force which made possible its own discovery and impressment.

[Here follow chapters which outline the chief applications of flame and of electricity.]

Let us compare electricity with its precursor, fire, and we shall understand the revolution by which fire is now in so many tasks supplanted by the electric pulse which, the while, creates for itself a thousand fields denied to flame. Copper is an excellent thermal conductor, and yet it transmits heat almost infinitely more slowly than it conveys electricity. One end of a thick copper rod ten feet long may be safely held in the hand while the other end is heated to redness, yet one millionth part of this same energy, if in the form of electricity, would traverse the rod in one 100,000,000th part of a second. Compare next electricity with light, often the companion of heat. Light travels in straight lines only; electricity can go round a corner every inch for miles, and, none the worse, yield a brilliant beam at the end of its journey. Indirectly, therefore, electricity enables us to conduct either heat or light as if both were flexible pencils of rays, and subject to but the smallest tolls in their travel.

We have remarked upon such methods as those of the electric welder which summon intense heat without fire, and we have glanced at the electric lamps which shine just because combustion is impossible through their rigid exclusion of air. Then for a moment we paused to look at the plating baths which have developed themselves into a commanding rivalry with the blaze of the smelting furnace, with the flame which from time immemorial has filled the ladle of the founder and moulder. Thus methods that commenced in dismissing flame end boldly by dispossessing heat itself. But, it may be said, this usurping electricity usually finds its source, after all, in combustion under a steam-boiler. True, but mark the harnessing of Niagara, of the Lachine Rapids near Montreal, of a thousand streams elsewhere. In the near future motive power of Nature's giving is to be wasted less and less, and perforce will more and more exclude heat from the chain of transformations which issue in the locomotive's flight, in the whirl of factory and mill. Thus in some degree is allayed the fear, never well grounded, that when the coal fields of the globe are spent civilization must collapse. As the electrician hears this foreboding he recalls how much fuel is wasted in converting heat into electricity. He looks beyond either turbine or shaft turned by wind or tide, and, remembering that the metal dissolved in his battery yields at his will its full content of energy, either as heat or electricity, he asks, Why may not coal or forest tree, which are but other kinds of fuel, be made to do the same?

One of the earliest uses of light was a means of communicating intelligence, and to this day the signal lamp and the red fire of the mariner are as useful as of old. But how much wider is the field of electricity as it creates the telegraph and the telephone! In the telegraph we have all that a pencil of light could be were it as long as an equatorial girdle and as flexible as a silken thread. In the telephone for nearly two thousand miles the pulsations of the speaker's voice are not only audible, but retain their characteristic tones.

In the field of mechanics electricity is decidedly preferable to any other agent. Heat may be transformed into motive power by a suitable engine, but there its adaptability is at an end. An electric current drives not only a motor, but every machine and tool attached to the motor, the whole executing tasks of a delicacy and complication new to industrial art. On an electric railroad an identical current propels the train, directs it by telegraph, operates its signals, provides it with light and heat, while it stands ready to give constant verbal communication with any station on the line, if this be desired.

In the home electricity has equal versatility, at once promoting healthfulness, refinement and safety. Its tiny button expels the hazardous match as it lights a lamp which sends forth no baleful fumes. An electric fan brings fresh air into the house--in summer as a grateful breeze. Simple telephones, quite effective for their few yards of wire, give a better because a more flexible service than speaking-tubes. Few invalids are too feeble to whisper at the light, portable ear of metal. Sewing-machines and the more exigent apparatus of the kitchen and laundry transfer their demands from flagging human muscles to the tireless sinews of electric motors--which ask no wages when they stand unemployed. Similar motors already enjoy favour in working the elevators of tall dwellings in cities. If a householder is timid about burglars, the electrician offers him a sleepless watchman in the guise of an automatic alarm; if he has a dread of fire, let him dispose on his walls an array of thermometers that at the very inception of a blaze will strike a gong at headquarters. But these, after all, are matters of minor importance in comparison with the foundations upon which may be reared, not a new piece of mechanism, but a new science or a new art.

In the recent swift subjugation of the territory open alike to the chemist and the electrician, where each advances the quicker for the other's company, we have fresh confirmation of an old truth--that the boundary lines which mark off one field of science from another are purely artificial, are set up only for temporary convenience. The chemist has only to dig deep enough to find that the physicist and himself occupy common ground. "Delve from the surface of your sphere to its heart, and at once your radius joins every other." Even the briefest glance at electro-chemistry should pause to acknowledge its profound debt to the new theories as to the bonding of atoms to form molecules, and of the continuity between solution and electrical dissociation. However much these hypotheses may be modified as more light is shed on the geometry and the journeyings of the molecule, they have for the time being recommended themselves as finder-thoughts of golden value. These speculations of the chemist carry him back perforce to the days of his childhood. As he then joined together his black and white bricks he found that he could build cubes of widely different patterns. It was in propounding a theory of molecular architecture that Kekul?gave an impetus to a vast and growing branch of chemical industry--that of the synthetic production of dyes and allied compounds.

It was in pure research, in paths undirected to the market-place, that such theories have been thought out. Let us consider electricity as an aid to investigation conducted for its own sake. The chief physical generalization of our time, and of all time, the persistence of force, emerged to view only with the dawn of electric art. When it was observed that electricity might become heat, light, chemical action, or mechanical motion, that in turn any of these might produce electricity, it was at once indicated that all these phases of energy might differ from each other only as the movements in circles, volutes, and spirals of ordinary mechanism. The suggestion was confirmed when electrical measurers were refined to the utmost precision, and a single quantum of energy was revealed a very Proteus in its disguises, yet beneath these disguises nothing but constancy itself.

"There is that scattereth, and yet increaseth; and there is that withholdeth more than is meet, but it tendeth to poverty." Because the geometers of old patiently explored the properties of the triangle, the circle, and the ellipse, simply for pure love of truth, they laid the corner-stones for the arts of the architect, the engineer, and the navigator. In like manner it was the disinterested work of investigation conducted by Ampere, Faraday, Henry and their compeers, in ascertaining the laws of electricity which made possible the telegraph, the telephone, the dynamo, and the electric furnace. The vital relations between pure research and economic gain have at last worked themselves clear. It is perfectly plain that a man who has it in him to discover laws of matter and energy does incomparably more for his kind than if he carried his talents to the mint for conversion into coin. The voyage of a Columbus may not immediately bear as much fruit as the uncoverings of a mine prospector, but in the long run a Columbus makes possible the finding many mines which without him no prospector would ever see. Therefore let the seed-corn of knowledge be planted rather than eaten. But in choosing between one research and another it is impossible to foretell which may prove the richer in its harvests; for instance, all attempts thus far economically to oxidize carbon for the production of electricity have failed, yet in observations that at first seemed equally barren have lain the hints to which we owe the incandescent lamp and the wireless telegraph.

Perhaps the most promising field of electrical research is that of discharges at high pressures; here the leading American investigators are Professor John Trowbridge and Professor Elihu Thomson. Employing a tension estimated at one and a half millions volts, Professor Trowbridge has produced flashes of lightning six feet in length in atmospheric air; in a tube exhausted to one-seventh of atmospheric pressure the flashes extended themselves to forty feet. According to this inquirer, the familiar rending of trees by lightning is due to the intense heat developed in an instant by the electric spark; the sudden expansion of air or steam in the cavities of the wood causes an explosion. The experiments of Professor Thomson confront him with some of the seeming contradictions which ever await the explorer of new scientific territory. In the atmosphere an electrical discharge is facilitated when a metallic terminal (as a lightning rod) is shaped as a point; under oil a point is the form least favourable to discharge. In the same line of paradox it is observed that oil steadily improves in its insulating effect the higher the electrical pressure committed to its keeping; with air as an insulator the contrary is the fact. These and a goodly array of similar puzzles will, without doubt, be cleared up as students in the twentieth century pass from the twilight of anomaly to the sunshine of ascertained law.

"Before there can be applied science there must be science to apply," and it is by enabling the investigator to know nature under a fresh aspect that electricity rises to its highest office. The laboratory routine of ascertaining the conductivity, polarisability, and other electrical properties of matter is dull and exacting work, but it opens to the student new windows through which to peer at the architecture of matter. That architecture, as it rises to his view, discloses one law of structure after another; what in a first and clouded glance seemed anomaly is now resolved and reconciled; order displays itself where once anarchy alone appeared. When the investigator now needs a substance of peculiar properties he knows where to find it, or has a hint for its creation--a creation perhaps new in the history of the world. As he thinks of the wealth of qualities possessed by his store of alloys, salts, acids, alkalies, new uses for them are borne into his mind. Yet more--a new orchestration of inquiry is possible by means of the instruments created for him by the electrician, through the advances in method which these instruments effect. With a second and more intimate point of view arrives a new trigonometry of the particle, a trigonometry inconceivable in pre-electric days. Hence a surround is in progress which early in the twentieth century may go full circle, making atom and molecule as obedient to the chemist as brick and stone are to the builder now.

The laboratory investigator and the commercial exploiter of his discoveries have been by turns borrower and lender, to the great profit of both. What Leyden jar could ever be constructed of the size and revealing power of an Atlantic cable? And how many refinements of measurement, of purification of metals, of precision in manufacture, have been imposed by the colossal investments in deep-sea telegraphy alone! When a current admitted to an ocean cable, such as that between Brest and New York, can choose for its path either 3,540 miles of copper wire or a quarter of an inch of gutta-percha, there is a dangerous opportunity for escape into the sea, unless the current is of nicely adjusted strength, and the insulator has been made and laid with the best-informed skill, the most conscientious care. In the constant tests required in laying the first cables Lord Kelvin (then Professor William Thomson) felt the need for better designed and more sensitive galvanometers or current measurers. His great skill both as a mathematician and a mechanician created the existing instruments, which seem beyond improvement. They serve not only in commerce and manufacture, but in promoting the strictly scientific work of the laboratory. Now that electricity purifies copper as fire cannot, the mathematician is able to treat his problems of long-distance transmission, of traction, of machine design, with an economy and certainty impossible when his materials were not simply impure, but impure in varying and indefinite degrees. The factory and the workshop originally took their magneto-machines from the experimental laboratory; they have returned them remodelled beyond recognition as dynamos and motors of almost ideal effectiveness.

A galvanometer actuated by a thermo-electric pile furnishes much the most sensitive means of detecting changes of temperature; hence electricity enables the physicist to study the phenomena of heat with new ease and precision. It was thus that Professor Tyndall conducted the classical researches set forth in his "Heat as a Mode of Motion," ascertaining the singular power to absorb terrestrial heat which makes the aqueous vapours of the atmosphere act as an indispensable blanket to the earth.

And how vastly has electricity, whether in the workshop or laboratory, enlarged our conceptions of the forces that thrill space, of the substances, seemingly so simple, that surround us--substances that propound questions of structure and behaviour that silence the acutest investigator. "You ask me," said a great physicist, "if I have a theory of the universe? Why, I haven't even a theory of magnetism!"

The conventional phrase "conducting a current" is now understood to be mere figure of speech; it is thought that a wire does little else than give direction to electric energy. Pulsations of high tension have been proved to be mainly superficial in their journeys, so that they are best conveyed (or convoyed) by conductors of tubular form. And what is it that moves when we speak of conduction? It seems to be now the molecule of atomic chemistry, and anon the same ether that undulates with light or radiant heat. Indeed, the conquest of electricity means so much because it impresses the molecule and the ether into service as its vehicles of communication. Instead of the old-time masses of metal, or bands of leather, which moved stiffly through ranges comparatively short, there is to-day employed a medium which may traverse 186,400 miles in a second, and with resistances most trivial in contrast with those of mechanical friction.

And what is friction in the last analysis but the production of motion in undesired forms, the allowing valuable energy to do useless work? In that amazing case of long distance transmission, common sunshine, a solar beam arrives at the earth from the sun not one whit the weaker for its excursion of 92,000,000 miles. It is highly probable that we are surrounded by similar cases of the total absence of friction in the phenomena of both physics and chemistry, and that art will come nearer and nearer to nature in this immunity is assured when we see how many steps in that direction have already been taken by the electrical engineer. In a preceding page a brief account was given of the theory that gases and vapours are in ceaseless motion. This motion suffers no abatement from friction, and hence we may infer that the molecules concerned are perfectly elastic. The opinion is gaining ground among physicists that all the properties of matter, transparency, chemical combinability, and the rest, are due to immanent motion in particular orbits, with diverse velocities. If this be established, then these motions also suffer no friction, and go on without resistance forever.

As the investigators in the vanguard of science discuss the constitution of matter, and weave hypotheses more or less fruitful as to the interplay of its forces, there is a growing faith that the day is at hand when the tie between electricity and gravitation will be unveiled--when the reason why matter has weight will cease to puzzle the thinker. Who can tell what relief of man's estate may be bound up with the ability to transform any phase of energy into any other without the circuitous methods and serious losses of to-day! In the sphere of economic progress one of the supreme advances was due to the invention of money, the providing a medium for which any salable thing may be exchanged, with which any purchasable thing may be bought. As soon as a shell, or a hide, or a bit of metal was recognized as having universal convertibility, all the delays and discounts of barter were at an end. In the world of physics and chemistry the corresponding medium is electricity; let it be produced as readily as it produces other modes of motion, and human art will take a stride forward such as when Volta disposed his zinc and silver discs together, or when Faraday set a magnet moving around a copper wire.

For all that the electric current is not as yet produced as economically as it should be, we do wrong if we regard it as an infant force. However much new knowledge may do with electricity in the laboratory, in the factory, or in the exchange, some of its best work is already done. It is not likely ever to perform a greater feat than placing all mankind within ear-shot of each other. Were electricity unmastered there could be no democratic government of the United States. To-day the drama of national affairs is more directly in view of every American citizen than, a century ago, the public business of Delaware could be to the men of that little State. And when on the broader stage of international politics misunderstandings arise, let us note how the telegraph has modified the hard-and-fast rules of old-time diplomacy. To-day, through the columns of the press, the facts in controversy are instantly published throughout the world, and thus so speedily give rise to authoritative comment that a severe strain is put upon negotiators whose tradition it is to be both secret and slow.

Railroads, with all they mean for civilization, could not have extended themselves without the telegraph to control them. And railroads and telegraphs are the sinews and nerves of national life, the prime agencies in welding the diverse and widely separated States and Territories of the Union. A Boston merchant builds a cotton-mill in Georgia; a New York capitalist opens a copper-mine in Arizona. The telegraph which informs them day by day how their investments prosper tells idle men where they can find work, where work can seek idle men. Chicago is laid in ashes, Charleston topples in earthquake, Johnstown is whelmed in flood, and instantly a continent springs to their relief. And what benefits issue in the strictly commercial uses of the telegraph! At its click both locomotive and steamship speed to the relief of famine in any quarter of the globe. In times of plenty or of dearth the markets of the globe are merged and are brought to every man's door. Not less striking is the neighbourhood guild of science, born, too, of the telegraph. The day after R 鍊 tgen announced his X rays, physicists on every continent were repeating his experiments--were applying his discovery to the healing of the wounded and diseased. Let an anti-toxin for diphtheria, consumption, or yellow fever be proposed, and a hundred investigators the world over bend their skill to confirm or disprove, as if the suggester dwelt next door.

On a stage less dramatic, or rather not dramatic at all, electricity works equal good. Its motor freeing us from dependence on the horse is spreading our towns and cities into their adjoining country. Field and garden compete with airless streets. The sunny cottage is in active rivalry with the odious tenement-house. It is found that transportation within the gates of a metropolis has an importance second only to the means of transit which links one city with another. The engineer is at last filling the gap which too long existed between the traction of horses and that of steam. In point of speed, cleanliness, and comfort such an electric subway as that of South London leaves nothing to be desired. Throughout America electric roads, at first suburban, are now fast joining town to town and city to city, while, as auxiliaries to steam railroads, they place sparsely settled communities in the arterial current of the world, and build up a ready market for the dairyman and the fruit-grower. In its saving of what Mr. Oscar T. Crosby has called "man-hours" the third-rail system is beginning to oust steam as a motive power from trunk-lines. Already shrewd railroad managers are granting partnerships to the electricians who might otherwise encroach upon their dividends. A service at first restricted to passengers has now extended itself to the carriage of letters and parcels, and begins to reach out for common freight. We may soon see the farmer's cry for good roads satisfied by good electric lines that will take his crops to market much more cheaply and quickly than horses and macadam ever did. In cities, electromobile cabs and vans steadily increase in numbers, furthering the quiet and cleanliness introduced by the trolley car.

A word has been said about the blessings which electricity promises to country folk, yet greater are the boons it stands ready to bestow in the hives of population. Until a few decades ago the water-supply of cities was a matter not of municipal but of individual enterprise; water was drawn in large part from wells here and there, from lines of piping laid in favoured localities, and always insufficient. Many an epidemic of typhoid fever was due to the contamination of a spring by a cesspool a few yards away. To-day a supply such as that of New York is abundant and cheap because it enters every house. Let a centralized electrical service enjoy a like privilege, and it will offer a current which is heat, light, chemical energy, or motive power, and all at a wage lower than that of any other servant. Unwittingly, then, the electrical engineer is a political reformer of high degree, for he puts a new premium upon ability and justice at the City Hall. His sole condition is that electricity shall be under control at once competent and honest. Let us hope that his plea, joined to others as weighty, may quicken the spirit of civic righteousness so that some of the richest fruits ever borne in the garden of science and art may not be proffered in vain. Flame, the old-time servant, is individual; electricity, its successor and heir, is collective. Flame sits upon the hearth and draws a family together; electricity, welling from a public source, may bind into a unit all the families of a vast city, because it makes the benefit of each the interest of all.

But not every promise brought forward in the name of the electrician has his assent or sanction. So much has been done by electricity, and so much more is plainly feasible, that a reflection of its triumphs has gilded many a baseless dream. One of these is that the cheap electric motor, by supply power at home, will break up the factory system, and bring back the domestic manufacturing of old days. But if this power cost nothing at all the gift would leave the factory unassailed; for we must remember that power is being steadily reduced in cost from year to year, so that in many industries it has but a minor place among the expenses of production. The strength and profit of the factory system lie in its assembling a wide variety of machines, the first delivering its product to the second for another step toward completion, and so on until a finished article is sent to the ware-room. It is this minute subdivision of labour, together with the saving and efficiency that inure to a business conducted on an immense scale under a single manager, that bids us believe that the factory has come to stay. To be sure, a weaver, a potter, or a lens-grinder of peculiar skill may thrive at his loom or wheel at home; but such a man is far from typical in modern manufacture. Besides, it is very questionable whether the lamentations over the home industries of the past do not ignore evil concomitants such as still linger in the home industries of the present--those of the sweater's den, for example.

This rapid survey of what electricity has done and may yet do--futile expectation dismissed--has shown it the creator of a thousand material resources, the perfector of that communication of things, of power, of thought, which in every prior stage of advancement has marked the successive lifts of humanity. It was much when the savage loaded a pack upon a horse or an ox instead of upon his own back; it was yet more when he could make a beacon-flare give news or warning to a whole country-side, instead of being limited to the messages which might be read in his waving hands. All that the modern engineer was able to do with steam for locomotion is raised to a higher plane by the advent of his new power, while the long-distance transmission of electrical energy is contracting the dimensions of the planet to a scale upon which its cataracts in the wilderness drive the spindles and looms of the factory town, or illuminate the thoroughfares of cities. Beyond and above all such services as these, electricity is the corner-stone of physical generalization, a revealer of truths impenetrable by any other ray.

The subjugation of fire has done much in giving man a new independence of nature, a mighty armoury against evil. In curtailing the most arduous and brutalizing forms of toil, electricity, that subtler kind of fire, carries this emancipation a long step further, and, meanwhile, bestows upon the poor many a luxury which but lately was the exclusive possession of the rich. In more closely binding up the good of the bee with the welfare of the hive, it is an educator and confirmer of every social bond. In so far as it proffers new help in the war on pain and disease it strengthens the confidence of man in an Order of Right and Happiness which for so many dreary ages has been a matter rather of hope than of vision. Are we not, then, justified in holding electricity to be a multiplier of faculty and insight, a means of dignifying mind and soul, unexampled since man first kindled fire and rejoiced?

We have traced how dexterity rose to fire-making, how fire-making led to the subjugation of electricity. Much of the most telling work of fire can be better done by its great successor, while electricity performs many tasks possible only to itself. Unwitting truth there was in the simple fable of the captive who let down a spider's film, that drew up a thread, which in turn brought up a rope--and freedom. It was in 1800 on the threshold of the nineteenth century, that Volta devised the first electric battery. In a hundred years the force then liberated has vitally interwoven itself with every art and science, bearing fruit not to be imagined even by men of the stature of Watt, Lavoisier, or Humboldt. Compare this rapid march of conquest with the slow adaptation, through age after age, of fire to cooking, smelting, tempering. Yet it was partly, perhaps mainly, because the use of fire had drawn out man's intelligence and cultivated his skill that he was ready in the fulness of time so quickly to seize upon electricity and subdue it.

Electricity is as legitimately the offspring of fire as fire of the simple knack in which one savage in ten thousand was richer than his fellows. The principle of permutation, suggested in both victories, interprets not only how vast empire is won by a new weapon of prime dignity; it explains why such empires are brought under rule with ever-accelerated pace. Every talent only pioneers the way for the richer talents which are born from it.

FOOTNOTES:

[5] Permutations are the various ways in which two or more different things may be arranged in a row, all the things appearing in each row. Permutations are readily illustrated with squares or cubes of different colours, with numbers, or letters.

Permutations of two elements, 1 and 2, are (1 x 2) two; 1, 2; 2, 1; or a, b; b, a. Of three elements the permutations are (1 x 2 x 3) six; 1, 2, 3; 1, 3, 2; 2, 1, 3; 2, 3, 1; 3, 1, 2; 3, 2, 1; or a, b, c; a, c, b; b, a, c; b, c, a; c, a, b; c, b, a. Of four elements the permutations are (1 x 2 x 3 x 4) twenty-four; of five elements, one hundred and twenty, and so on. A new element or permutator multiplies by an increasing figure all the permutations it finds.

[6] Some years ago I sent an outline of this argument to Herbert Spencer, who replied: "I recognize a novelty and value in your inference that the law implies an increasing width of gap between lower and higher types as evolution advances."

COUNT RUMFORD IDENTIFIES HEAT WITH MOTION.

[Benjamin Thompson, who received the title of Count Rumford from the Elector of Bavaria, was born in Woburn, Massachusetts, in 1753. When thirty-one years of age he settled in Munich, where he devoted his remarkable abilities to the public service. Twelve years afterward he removed to England; in 1800 he founded the Royal Institution of London, since famous as the theatre of the labours of Davy, Faraday, Tyndall, and Dewar. He bequeathed to Harvard University a fund to endow a professorship of the application of science to the art of living: he instituted a prize to be awarded by the American Academy of Sciences for the most important discoveries and improvements relating to heat and light. In 1804 he married the widow of the illustrious chemist Lavoisier: he died in 1814. Count Rumford on January 25, 1798, read a paper before the Royal Society entitled "An Enquiry Concerning the Source of Heat Which Is Excited by Friction." The experiments therein detailed proved that heat is identical with motion, as against the notion that heat is matter. He thus laid the corner-stone of the modern theory that heat light, electricity, magnetism, chemical action, and all other forms of energy are in essence motion, are convertible into one another, and as motion are indestructible. The following abstract of Count Rumford's paper is taken from "Heat as a Mode of Motion," by Professor John Tyndall, published by D. Appleton & Co., New York. This work and "The Correlation and Conservation of Forces," edited by Dr. E. L. Youmans, published by the same house, will serve as a capital introduction to the modern theory that energy is motion which, however varied in its forms, is changeless in its quantity.]

Being engaged in superintending the boring of cannon in the workshops of the military arsenal at Munich, Count Rumford was struck with the very considerable degree of heat which a brass gun acquires, in a short time, in being bored, and with the still more intense heat (much greater than that of boiling water) of the metallic chips separated from it by the borer, he proposed to himself the following questions:

"Whence comes the heat actually produced in the mechanical operations above mentioned?

"Is it furnished by the metallic chips which are separated from the metal?"

If this were the case, then the capacity for heat of the parts of the metal so reduced to chips ought not only to be changed, but the change undergone by them should be sufficiently great to account for all the heat produced. No such change, however, had taken place, for the chips were found to have the same capacity as slices of the same metal cut by a fine saw, where heating was avoided. Hence, it is evident, that the heat produced could not possibly have been furnished at the expense of the latent heat of the metallic chips. Rumford describes these experiments at length, and they are conclusive.

He then designed a cylinder for the express purpose of generating heat by friction, by having a blunt borer forced against its solid bottom, while the cylinder was turned around its axis by the force of horses. To measure the heat developed, a small round hole was bored in the cylinder for the purpose of introducing a small mercurial thermometer. The weight of the cylinder was 113.13 pounds avoirdupois.

The borer was a flat piece of hardened steel, 0.63 of an inch thick, four inches long, and nearly as wide as the cavity of the bore of the cylinder, namely, three and one-half inches. The area of the surface by which its end was in contact with the bottom of the bore was nearly two and one-half inches. At the beginning of the experiment the temperature of the air in the shade, and also that of the cylinder, was 60?Fahr. At the end of thirty minutes, and after the cylinder had made 960 revolutions round its axis, the temperature was found to be 130?

Having taken away the borer, he now removed the metallic dust, or rather scaly matter, which had been detached from the bottom of the cylinder by the blunt steel borer, and found its weight to be 837 grains troy. "Is it possible," he exclaims, "that the very considerable quantity of heat produced in this experiment--a quantity which actually raised the temperature of above 113 pounds of gun-metal at least 70?of Fahrenheit's thermometer--could have been furnished by so inconsiderable a quantity of metallic dust and this merely in consequence of a change in its capacity of heat?"

"But without insisting on the improbability of this supposition, we have only to recollect that from the results of actual and decisive experiments, made for the express purpose of ascertaining that fact, the capacity for heat for the metal of which great guns are cast is not sensibly changed by being reduced to the form of metallic chips, and there does not seem to be any reason to think that it can be much changed, if it be changed at all, in being reduced to much smaller pieces by a borer which is less sharp."

He next surrounded his cylinder by an oblong deal-box, in such a manner that the cylinder could turn water-tight in the centre of the box, while the borer was pressed against the bottom of the cylinder. The box was filled with water until the entire cylinder was covered, and then the apparatus was set in action. The temperature of the water on commencing was 60?

"The result of this beautiful experiment," writes Rumford, "was very striking, and the pleasure it afforded me amply repaid me for all the trouble I had had in contriving and arranging the complicated machinery used in making it. The cylinder had been in motion but a short time, when I perceived, by putting my hand into the water, and touching the outside of the cylinder, that heat was generated.

"At the end of one hour the fluid, which weighed 18.77 pounds, or two and one-half gallons, had its temperature raised forty-seven degrees, being now 107?

"In thirty minutes more, or one hour and thirty minutes after the machinery had been set in motion, the heat of the water was 142?

"At the end of two hours from the beginning, the temperature was 178?

"At two hours and twenty minutes it was 200? and at two hours and thirty minutes it actually boiled!"

"It would be difficult to describe the surprise and astonishment expressed in the countenances of the bystanders on seeing so large a quantity of water heated, and actually made to boil, without any fire. Though, there was nothing that could be considered very surprising in this matter, yet I acknowledge fairly that it afforded me a degree of childish pleasure which, were I ambitious of the reputation of a grave philosopher, I ought most certainly rather to hide than to discover."

He then carefully estimates the quantity of heat possessed by each portion of his apparatus at the conclusion of the experiment, and, adding all together, finds a total sufficient to raise 26.58 pounds of ice-cold water to its boiling point, or through 180?Fahrenheit. By careful calculation, he finds this heat equal to that given out by the combustion of 2,303.8 grains (equal to four and eight-tenths ounces troy) of wax.

He then determines the "celerity" with which the heat was generated, summing up thus: "From the results of these computations, it appears that the quantity of heat produced equably, or in a continuous stream, if I may use the expression, by the friction of the blunt steel borer against the bottom of the hollow metallic cylinder, was greater than that produced in the combustion of nine wax-candles, each three-quarters of an inch in diameter, all burning together with clear bright flames.

"One horse would have been equal to the work performed, though two were actually employed. Heat may thus be produced merely by the strength of a horse, and, in a case of necessity, this heat might be used in cooking victuals. But no circumstances could be imagined in which this method of procuring heat would be advantageous, for more heat might be obtained by using the fodder necessary for the support of a horse as fuel."

[This is an extremely significant passage, intimating as it does, that Rumford saw clearly that the force of animals was derived from the food; no creation of force taking place in the animal body.]

"By meditating on the results of all these experiments, we are naturally brought to that great question which has so often been the subject of speculation among philosophers, namely, What is heat--is there any such thing as an igneous fluid? Is there anything that, with propriety, can be called caloric?

"We have seen that a very considerable quantity of heat may be excited by the friction of two metallic surfaces, and given off in a constant stream or flux in all directions, without interruption or intermission, and without any signs of diminution or exhaustion. In reasoning on this subject we must not forget that most remarkable circumstance, that the source of the heat generated by friction in these experiments appeared evidently to be inexhaustible. [The italics are Rumford's.] It is hardly necessary to add, that anything which any insulated body or system of bodies can continue to furnish without limitation cannot possibly be a material substance; and it appears to me to be extremely difficult, if not quite impossible, to form any distinct idea of anything capable of being excited and communicated in those experiments, except it be MOTION."

When the history of the dynamical theory of heat is written, the man who, in opposition to the scientific belief of his time, could experiment and reason upon experiment, as Rumford did in the investigation here referred to, cannot be lightly passed over. Hardly anything more powerful against the materiality of heat has been since adduced, hardly anything more conclusive in the way of establishing that heat is, what Rumford considered it to be, Motion.

VICTORY OF THE "ROCKET" LOCOMOTIVE.

[Part of Chapter XII. Part II, of "The Life of George Stephenson and of His Son, Robert Stephenson," by Samuel Smiles New York, Harper & Brothers, 1868.]

The works of the Liverpool and Manchester Railway were now approaching completion. But, strange to say, the directors had not yet decided as to the tractive power to be employed in working the line when open for traffic. The differences of opinion among them were so great as apparently to be irreconcilable. It was necessary, however, that they should, come to some decision without further loss of time, and many board meetings were accordingly held to discuss the subject. The old-fashioned and well-tried system of horse-haulage was not without its advocates; but, looking at the large amount of traffic which there was to be conveyed, and at the probable delay in the transit from station to station if this method were adopted, the directors, after a visit made by them to the Northumberland and Durham railways in 1828, came to the conclusion that the employment of horse-power was inadmissible.

Fixed engines had many advocates; the locomotive very few: it stood as yet almost in a minority of one--George Stephenson....

In the meantime the discussion proceeded as to the kind of power to be permanently employed for the working of the railway. The directors were inundated with schemes of all sorts for facilitating locomotion. The projectors of England, France, and America seemed to be let loose upon them. There were plans for working the waggons along the line by water-power. Some proposed hydrogen, and others carbonic acid gas. Atmospheric pressure had its eager advocates. And various kinds of fixed and locomotive steam-power were suggested. Thomas Gray urged his plan of a greased road with cog-rails; and Messrs. Vignolles and Ericsson recommended the adoption of a central friction-rail, against which two horizontal rollers under the locomotive, pressing upon the sides of this rail, were to afford the means of ascending the inclined planes....

The two best practical engineers of the day concurred in reporting substantially in favour of the employment of fixed engines. Not a single professional man of eminence could be found to coincide with the engineer of the railway in his preference for locomotive over fixed engine power. He had scarcely a supporter, and the locomotive system seemed on the eve of being abandoned. Still he did not despair. With the profession against him, and public opinion against him--for the most frightful stories went abroad respecting the dangers, the unsightliness, and the nuisance which the locomotive would create--Stephenson held to his purpose. Even in this, apparently the darkest hour of the locomotive, he did not hesitate to declare that locomotive railroads would, before many years had passed, be "the great highways of the world."

He urged his views upon the directors in all ways, in season, and, as some of them thought, out of season. He pointed out the greater convenience of locomotive power for the purposes of a public highway, likening it to a series of short unconnected chains, any one of which could be removed and another substituted without interruption to the traffic; whereas the fixed-engine system might be regarded in the light of a continuous chain extending between the two termini, the failure of any link of which would derange the whole. But the fixed engine party was very strong at the board, and, led by Mr. Cropper, they urged the propriety of forthwith adopting the report of Messrs. Walker and Rastrick. Mr. Sandars and Mr. William Rathbone, on the other hand, desired that a fair trial should be given to the locomotive; and they with reason objected to the expenditure of the large capital necessary to construct the proposed engine-houses, with their fixed engines, ropes, and machinery, until they had tested the powers of the locomotive as recommended by their own engineer. George Stephenson continued to urge upon them that the locomotive was yet capable of great improvements, if proper inducements were held out to inventors and machinists to make them; and he pledged himself that, if time were given him, he would construct an engine that should satisfy their requirements, and prove itself capable of working heavy loads along the railway with speed, regularity, and safety. At length, influenced by his persistent earnestness not less than by his arguments, the directors, at the suggestion of Mr. Harrison, determined to offer a prize of ?00 for the best locomotive engine, which, on a certain day, should be produced on the railway, and perform certain specified conditions in the most satisfactory manner.[7]

The requirements of the directors as to speed were not excessive. All that they asked for was that ten miles an hour should be maintained. Perhaps they had in mind the animadversions of the Quarterly Review on the absurdity of travelling at a greater velocity, and also the remarks published by Mr. Nicholas Wood, whom they selected to be one of the judges of the competition, in conjunction, with Mr. Rastrick, of Stourbridge, and Mr. Kennedy, of Manchester.

It was now felt that the fate of railways in a great measure depended upon the issue of this appeal to the mechanical genius of England. When the advertisement of the prize for the best locomotive was published, scientific men began more particularly to direct their attention to the new power which was thus struggling into existence. In the meantime public opinion on the subject of railway working remained suspended, and the progress of the undertaking was watched with intense interest.

During the progress of this important controversy with reference to the kind of power to be employed in working the railway, George Stephenson was in constant communication with his son Robert, who made frequent visits to Liverpool for the purpose of assisting his father in the preparation of his reports to the board on the subject. Mr. Swanwick remembers the vivid interest of the evening discussions which then took place between father and son as to the best mode of increasing the powers and perfecting the mechanism of the locomotive. He wondered at their quick perception and rapid judgment on each other's suggestions; at the mechanical difficulties which they anticipated and provided for in the practical arrangement of the machine; and he speaks of these evenings as most interesting displays of two actively ingenious and able minds stimulating each other to feats of mechanical invention, by which it was ordained that the locomotive engine should become what it now is. These discussions became more frequent, and still more interesting, after the public prize had been offered for the best locomotive by the directors of the railway, and the working plans of the engine which they proposed to construct had to be settled.

One of the most important considerations in the new engine was the arrangement of the boiler, and the extension of its heating surface to enable steam enough to be raised rapidly and continuously for the purpose of maintaining high rates of speed--the effect of high pressure engines being ascertained to depend mainly upon the quantity of steam which the boiler can generate, and upon its degree of elasticity when produced. The quantity of steam so generated, it will be obvious, must chiefly depend upon the quantity of fuel consumed in the furnace, and, by necessary consequence, upon the high rate of temperature maintained there.

It will be remembered that in Stephenson's first Killingworth engines he invited and applied the ingenious method of stimulating combustion in the furnace by throwing the waste steam into the chimney after performing its office in the cylinders, thereby accelerating the ascent of the current of air, greatly increasing the draught, and consequently the temperature of the fire. This plan was adopted by him, as we have seen, as early as 1815, and it was so successful that he himself attributed to it the greater economy of the locomotive as compared with horse-power. Hence the continuance of its use upon the Killingworth Railway.

Though the adoption of the steam blast greatly quickened combustion and contributed to the rapid production of high-pressure steam, the limited amount of heating surface presented to the fire was still felt to be an obstacle to the complete success of the locomotive engine. Mr. Stephenson endeavoured to overcome this by lengthening the boilers and increasing the surface presented by the flue-tubes. The "Lancashire Witch," which he built for the Bolton and Leigh Railway, and used in forming the Liverpool and Manchester Railway embankments, was constructed with a double tube, each of which contained a fire, and passed longitudinally through the boiler. But this arrangement necessarily led to a considerable increase in the weight of those engines, which amounted to about twelve tons each; and as six tons was the limit allowed for engines admitted to the Liverpool competition, it was clear that the time was come when the Killingworth engine must undergo a farther important modification.

For many years previous to this period, ingenious mechanics had been engaged in attempting to solve the problem of the best and most economical boiler for the production of high-pressure steam.

The use of tubes in boilers for increasing the heating surface had long been known. As early as 1780, Matthew Boulton employed copper tubes longitudinally in the boiler of the Wheal Busy engine in Cornwall--the fire passing through the tubes--and it was found that the production of steam was thereby considerably increased. The use of tubular boilers afterwards became common in Cornwall. In 1803, Woolf, the Cornish engineer, patented a boiler with tubes, with the same object of increasing the heating surface. The water was inside the tubes, and the fire of the boiler outside. Similar expedients were proposed by other inventors. In 1815 Trevithick invented his light high-pressure boiler for portable purposes, in which, to "expose a large surface to the fire," he constructed the boiler of a number of small perpendicular tubes "opening into a common reservoir at the top." In 1823 W. H. James contrived a boiler composed of a series of annular wrought-iron tubes, placed side by side and bolted together, so as to form by their union a long cylindrical boiler, in the centre of which, at the end, the fireplace was situated. The fire played round the tubes, which contained the water. In 1826 James Neville took out a patent for a boiler with vertical tubes surrounded by the water, through which the heated air of the furnace passed, explaining also in his specification that the tubes might be horizontal or inclined, according to circumstances. Mr. Goldsworthy, the persevering adaptor of steam-carriages to travelling on common roads, applied the tubular principle in the boiler of his engine, in which the steam was generated within the tubes; while the boiler invented by Messrs. Summer and Ogle for their turnpike-road steam-carriage consisted of a series of tubes placed vertically over the furnace, through which the heated air passed before reaching the chimney.

About the same time George Stephenson was trying the effect of introducing small tubes in the boilers of his locomotives, with the object of increasing their evaporative power. Thus, in 1829, he sent to France two engines constructed at the Newcastle works for the Lyons and St. Etienne Railway, in the boilers of which tubes were placed containing water. The heating surface was thus considerably increased; but the expedient was not successful, for the tubes, becoming furred with deposit, shortly burned out and were removed. It was then that M. Seguin, the engineer of the railway, pursuing the same idea, is said to have adopted his plan of employing horizontal tubes through which the heated air passed in streamlets, and for which he took out a French patent.

In the meantime Mr. Henry Booth, secretary to the Liverpool and Manchester Railway, whose attention had been directed to the subject on the prize being offered for the best locomotive to work that line, proposed the same method, which, unknown to him, Matthew Boulton had employed but not patented, in 1780, and James Neville had patented, but not employed, in 1826; and it was carried into effect by Robert Stephenson in the construction of the "Rocket," which won the prize at Rainhill in October, 1829. The following is Mr. Booth's account in a letter to the author:

"I was in almost daily communication with Mr. Stephenson at the time, and I was not aware that he had any intention of competing for the prize till I communicated to him my scheme of a multitubular boiler. This new plan of boiler comprised the introduction of numerous small tubes, two or three inches in diameter, and less than one-eighth of an inch thick, through which to carry the fire instead of a single tube or flue eighteen inches in diameter, and about half an inch thick, by which plan we not only obtain a very much larger heating surface, but the heating surface is much more effective, as there intervenes between the fire and the water only a thin sheet of copper or brass, not an eighth of an inch thick, instead of a plate of iron of four times the substance, as well as an inferior conductor of heat.

"When the conditions of trial were published, I communicated my multitubular plan to Mr. Stephenson, and proposed to him that we should jointly construct an engine and compete for the prize. Mr. Stephenson approved the plan, and agreed to my proposal. He settled the mode in which the fire-box and tubes were to be mutually arranged and connected, and the engine was constructed at the works of Messrs. Robert Stephenson & Co., Newcastle-on-Tyne.

"I am ignorant of M. Seguin's proceedings in France, but I claim to be the inventor in England, and feel warranted in stating, without reservation, that until I named my plan to Mr. Stephenson, with a view to compete for the prize at Rainhill, it had not been tried, and was not known in this country."

From the well-known high character of Mr. Booth, we believe his statement to be made in perfect good faith, and that he was as much in ignorance of the plan patented by Neville as he was of that of Seguin. As we have seen, from the many plans of tubular boilers invented during the preceding thirty years, the idea was not by any means new; and we believe Mr. Booth to be entitled to the merit of inventing the method by which the multitubular principle was so effectually applied in the construction of the famous "Rocket" engine.

The principal circumstances connected with the construction of the "Rocket," as described by Robert Stephenson to the author, may be briefly stated. The tubular principle was adopted in a more complete manner than had yet been attempted. Twenty-five copper tubes, each three inches in diameter, extended from one end of the boiler to the other, the heated air passing through them on its way to the chimney; and the tubes being surrounded by the water of the boiler, it will be obvious that a large extension of the heating surface was thus effectually secured. The principal difficulty was in fitting the copper tubes in the boiler ends so as to prevent leakage. They were manufactured by a Newcastle coppersmith, and soldered to brass screws which were screwed into the boiler ends, standing out in great knobs. When the tubes were thus fitted, and the boiler was filled with water, hydraulic pressure was applied; but the water squirted out at every joint, and the factory floor was soon flooded. Robert went home in despair; and in the first moment of grief he wrote to his father that the whole thing was a failure. By return of post came a letter from his father, telling him that despair was not to be thought of--that he must "try again;" and he suggested a mode of overcoming the difficulty, which his son had already anticipated and proceeded to adopt. It was, to bore clean holes in the boiler ends, fit in the smooth copper tubes as tightly as possible, solder up, and then raise the steam. This plan succeeded perfectly, the expansion of the copper tubes completely filling up all interstices, and producing a perfectly water-tight boiler, capable of withstanding extreme external pressure.

The mode of employing the steam-blast for the purpose of increasing the draught in the chimney was also the subject of numerous experiments. When the engine was first tried, it was thought that the blast in the chimney was not sufficiently strong for the purpose of keeping up the intensity of fire in the furnace, so as to produce high-pressure steam with the required velocity. The expedient was therefore adopted of hammering the copper tubes at the point at which they entered the chimney, whereby the blast was considerably sharpened; and on a farther trial it was found that the draught was increased to such an extent as to enable abundance of steam to be raised. The rationale of the blast may be simply explained by referring to the effect of contracting the pipe of a water-hose, by which the force of the jet of water is proportionately increased. Widen the nozzle of the pipe, and the jet is in like manner diminished. So it is with the steam-blast in the chimney of the locomotive.

Doubts were, however, expressed whether the greater draught obtained by the contraction of the blast-pipe was not counterbalanced in some degree by the negative pressure upon the piston. Hence a series of experiments was made with pipes of different diameters, and their efficiency was tested by the amount of vacuum that was produced in the smoke-box. The degree of rarefaction was determined by a glass tube fixed to the bottom of the smoke-box and descending into a bucket of water, the tube being open at both ends. As the rarefaction took place, the water would, of course, rise in the tube, and the height to which it rose above the surface of the water in the bucket was made the measure of the amount of rarefaction. These experiments proved that a considerable increase of draught was obtained by the contraction of the orifice; accordingly, the two blast-pipes opening from the cylinders into either side of the "Rocket" chimney, and turned up within it, were contracted slightly below the area of the steam-ports, and before the engine left the factory, the water rose in the glass tube three inches above the water in the bucket.

The other arrangements of the "Rocket" were briefly these: the boiler was cylindrical, with flat ends, six feet in length, and three feet four inches in diameter. The upper half of the boiler was used as a reservoir for the steam, the lower half being filled with water. Through the lower part the copper tubes extended, being open to the fire-box at one end, and to the chimney at the other. The fire-box, or furnace, two feet wide and three feet high, was attached immediately behind the boiler, and was also surrounded with water. The cylinders of the engine were placed on each side of the boiler, in an oblique position, one end being nearly level with the top of the boiler at its after end, and the other pointing toward the centre of the foremost or driving pair of wheels, with which the connection was directly made from the piston-rod to a pin on the outside of the wheel. The engine, together with its load of water, weighed only four tons and a quarter; and it was supported on four wheels, not coupled. The tender was four-wheeled, and similar in shape to a waggon--the foremost part holding the fuel, and the hind part a water cask.

When the "Rocket" was finished it was placed upon the Killingworth Railway for the purpose of experiment. The new boiler arrangement was found perfectly successful. The steam was raised rapidly and continuously, and in a quantity which then appeared marvellous. The same evening Robert despatched a letter to his father at Liverpool, informing him, to his great joy, that the "Rocket" was "all right," and would be in complete working trim by the day of trial. The engine was shortly after sent by waggon to Carlisle, and thence shipped for Liverpool.

The time so much longed for by George Stephenson had now arrived, when the merits of the passenger locomotive were about to be put to the test. He had fought the battle for it until now almost single-handed. Engrossed by his daily labours and anxieties, and harassed by difficulties and discouragements which would have crushed the spirit of a less resolute man, he had held firmly to his purpose through good and through evil report. The hostility which he experienced from some of the directors opposed to the adoption of the locomotive was the circumstance that caused him the greatest grief of all; for where he had looked for encouragement, he found only carping and opposition. But his pluck never failed him; and now the "Rocket" was upon the ground to prove, to use his own words, "whether he was a man of his word or not."

On the day appointed for the great competition of locomotives at Rainhill the following engines were entered for the prize:

1. Messrs. Braithwaite and Fricsson's "Novelty."

2. Mr. Timothy Hackworth's "Sanspareil."

3. Messrs. R. Stephenson & Co.'s "Rocket."

4. Mr. Burstall's "Perseverance."

The ground on which the engines were to be tried was a level piece of railroad, about two miles in length. Each was required to make twenty trips, or equal to a journey of seventy miles, in the course of the day, and the average rate of travelling was to be not under ten miles an hour. It was determined that, to avoid confusion, each engine should be tried separately, and on different days.

The day fixed for the competition was the 1st of October, but, to allow sufficient time to get the locomotives into good working order, the directors extended it to the 6th. It was quite characteristic of the Stephensons that, although their engine did not stand first on the list for trial, it was the first that was ready, and it was accordingly ordered out by the judges for an experimental trip. Yet the "Rocket" was by no means the "favourite" with either the judges or the spectators. Nicholas Wood has since stated that the majority of the judges were strongly predisposed in favour of the "Novelty," and that "nine-tenths, if not ten-tenths, of the persons present were against the "Rocket" because of its appearance." Nearly every person favoured some other engine, so that there was nothing for the "Rocket" but the practical test. The first trip made by it was quite successful. It ran about twelve miles, without interruption, in about fifty-three minutes.

The "Novelty" was next called out. It was a light engine, very compact in appearance, carrying the water and fuel upon the same wheels as the engine. The weight of the whole was only three tons and one hundred-weight. A peculiarity of this engine was that the air was driven or forced through the fire by means of bellows. The day being now far advanced, and some dispute having arisen as to the method of assigning the proper load for the "Novelty," no particular experiment was made further than that the engine traversed the line by way of exhibition, occasionally moving at the rate of twenty-four miles an hour. The "Sanspareil," constructed by Mr. Timothy Hackworth, was next exhibited, but no particular experiment was made with it on this day. This engine differed but little in its construction from the locomotive last supplied by the Stephensons to the Stockton and Darlington Railway, of which Mr. Hackworth was the locomotive foreman.

The contest was postponed until the following day; but, before the judges arrived on the ground, the bellows for creating the blast in the "Novelty" gave way, and it was found incapable of going through its performance. A defect was also detected in the boiler of the "Sanspareil," and some further time was allowed to get it repaired. The large number of spectators who had assembled to witness the contest were greatly disappointed at this postponement; but, to lessen it, Stephenson again brought out the "Rocket," and, attaching it to a coach containing thirty persons, he ran them along the line at a rate of from twenty-four to thirty miles an hour, much to their gratification and amazement. Before separating, the judges ordered the engine to be in readiness by eight o'clock on the following morning, to go through its definite trial according to the prescribed conditions.

On the morning of the 8th of October the "Rocket" was again ready for the contest. The engine was taken to the extremity of the stage, the fire-box was filled with coke, the fire lighted, and the steam raised until it lifted the safety-valve loaded to a pressure of fifty pounds to the square inch. This proceeding occupied fifty-seven minutes. The engine then started on its journey, dragging after it about thirteen tons' weight in waggons, and made the first ten trips backward and forward along two miles of road, running the thirty-five miles, including stoppages, in an hour and forty-eight minutes. The second ten trips were in like manner performed in two hours and three minutes. The maximum velocity attained during the trial trip was twenty-nine miles an hour, or about three times the speed that one of the judges of the competition had declared to be the limit of possibility. The average speed at which the whole of the journeys was performed was fifteen miles an hour, or five miles beyond the rate specified in the conditions published by the company. The entire performance excited the greatest astonishment among the assembled spectators; the directors felt confident that their enterprise was now on the eve of success; and George Stephenson rejoiced to think that, in spite of all false prophets and fickle counsellors, the locomotive system was now safe. When the "Rocket," having performed all the conditions of the contest, arrived at the "grand stand" at the close of its day's successful run, Mr. Cropper--one of the directors favourable to the fixed engine system--lifted up his hands, and exclaimed, "Now has George Stephenson at last delivered himself...."

The "Rocket" had eclipsed the performance of all locomotive engines that had yet been constructed, and outstripped even the sanguine expectations of its constructors. It satisfactorily answered the report of Messrs. Walker and Rastrick, and established the efficiency of the locomotive for working the Liverpool and Manchester Railway, and, indeed, all future railways. The "Rocket" showed that a new power had been born into the world, full of activity and strength, with boundless capability of work. It was the simple but admirable contrivance of the steam-blast, and its combination with the multitubular boiler, that at once gave locomotion a vigorous life, and secured the triumph of the railway system.[8]

FOOTNOTES:

[7] The conditions were these:

1. The engine must effectually consume its own smoke.

2. The engine, if of six tons' weight, must be able to draw after it, day by day, twenty tons' weight (including the tender and water-tank) at ten miles an hour, with a pressure of steam on the boiler not exceeding fifty pounds to the square inch.

3. The boiler must have two safety-valves, neither of which must be fastened down, and one of them be completely out of the control of the engine-man.

4. The engine and boiler must be supported on springs, and rest on six wheels, the height of the whole not exceeding fifteen feet to the top of the chimney.

5. The engine, with water, must not weigh more than six tons; but an engine of less weight would be preferred on its drawing a proportionate load behind it; if of only four and a half tons, then it might be put on only four wheels. The company will be at liberty to test the boiler, etc., by a pressure of one hundred and fifty pounds to the square inch.

6. A mercurial gauge must be affixed to the machine, showing the steam pressure above forty-five pounds per square inch.

7. The engine must be delivered, complete and ready for trial, at the Liverpool end of the railway, not later than the 1st of October, 1829.

8. The price of the engine must not exceed ?50.

Many persons of influence declared the conditions published by the directors of the railway chimerical in the extreme. One gentleman of some eminence in Liverpool, Mr. P. Ewart, who afterward filled the office of Government Inspector of Post-office Steam Packets, declared that only a parcel of charlatans would ever have issued such a set of conditions; that it had been proved to be impossible to make a locomotive engine go at ten miles an hour; but if it ever was done, he would undertake to eat a stewed engine-wheel for his breakfast.

[8] When heavier and more powerful engines were brought upon the road, the old "Rocket," becoming regarded as a thing of no value, was sold in 1837. It has since been transferred to the Museum of Patents at South Kensington, London, where it is still to be seen.

###

www.ingramcontent.com/pod-product-compliance
Lightning Source LLC
Chambersburg PA
CBHW070909180526
45168CB00005B/1980